Medical Mic

2.00
(1997)

LECTURE NOTES ON

Medical Microbiology

TOM ELLIOTT
MB BS BTech BMedSci PhD FRMS FRCPath
Consultant Microbiologist
Department of Clinical Microbiology
University Hospital NHS Trust
Birmingham
Professor Associate
Department of Biology and Biochemistry
Brunel University
Honorary Senior Lecturer
University of Birmingham

MARK HASTINGS
MD MRCPath
Consultant Microbiologist
Department of Clinical Microbiology
University Hospital NHS Trust
Birmingham

ULRICH DESSELBERGER
MD FRCP(Glas, Lon) FRCPath
Consultant Virologist
Clinical Microbiology and Public Health Laboratory
Addenbrooke's Hospital
Cambridge

Third edition

b
**Blackwell
Science**

© 1975, 1978, 1997 by
Blackwell Science Ltd
Editorial Offices:
Osney Mead, Oxford OX2 0EL
25 John Street, London WC1N 2BL
23 Ainslie Place, Edinburgh EH3 6AJ
350 Main Street, Malden
 MA 02148 5018, USA
54 University Street, Carlton
 Victoria 3053, Australia
10, rue Casimir Delavigne
 75006 Paris, France

Other Editorial Offices:
Blackwell Wissenschafts-Verlag GmbH
Kurfürstendamm 57
10707 Berlin, Germany

Blackwell Science KK
MG Kodenmacho Building
7–10 Kodenmacho Nihombashi
Chuo-ku, Tokyo 104, Japan

The right of the Author to be
identified as the Author of this Work
has been asserted in accordance
with the Copyright, Designs and
Patents Act 1988.

All rights reserved. No part of
this publication may be reproduced,
stored in a retrieval system, or
transmitted, in any form or by any
means, electronic, mechanical,
photocopying, recording or otherwise,
except as permitted by the UK
Copyright, Designs and Patents Act
1988, without the prior permission
of the copyright owner.

First published (as Lecture Notes on Bacteriology) 1967
First edition 1975
Second edition 1978
Reprinted 1979, 1983, 1986
Third edition 1997
Reprinted 1998
International edition 1997

Set by Excel Typesetters Co., Hong Kong
Printed and bound in Great Britain
at the University Press, Cambridge

The Blackwell Science logo is a
trade mark of Blackwell Science Ltd,
registered at the United Kingdom
Trade Marks Registry

DISTRIBUTORS

Marston Book Services Ltd
PO Box 269
Abingdon
Oxford OX14 4YN
(*Orders*: Tel: 01235 465500
 Fax: 01235 465555)

USA
Blackwell Science, Inc.
Commerce Place
350 Main Street
Malden, MA 02148 5018
(*Orders*: Tel: 800 759 6102
 781 388 8250
 Fax: 781 388 8255)

Canada
Login Brothers Book Company
324 Saulteaux Crescent
Winnipeg, Manitoba R3J 3T2
(*Orders*: Tel: 204 224-4068)

Australia
Blackwell Science Pty Ltd
54 University Street
Carlton, Victoria 3053
(*Orders*: Tel: 03 9347 0300
 Fax: 03 9347 5001)

A catalogue record for this title
is available from the British Library

ISBN 0-632-02446-1 (BSL)
ISBN 0-632-03534-X (International edition)

Library of Congress
Cataloging-in-publication Data

Elliott, Thomas S.J.
 Lecture notes on medical microbiology. –3rd ed.
 / Thomas S.J. Elliott, Mark Hastings, Ulrich
 Desselberger.
  p. cm.
 Rev. ed. of: Lecture notes on medical microbiology
 / R.R. Gillies. 2nd ed. c1978.
 Includes bibliographical references and index.
 ISBN 0-632-02446-1. – ISBN 0-632-03534-X
 (International edition)
 I. Medical microbiology – Outlines, syllabi, etc.
 I. Hastings, Mark. II. Desselberger, U.
 III. Gillies, R. R. (Robert Reid)
 Lecture notes on medical microbiology. IV. Title
  [DNLM: I. Microbiology. 2. Communicable
 Diseases – microbiology.
 QW 4 E46L 1996]
 QR46.G49 1996
 616'.01–dc20
 DNLM/DLC
 for Library of Congress   96-2259
                                         CIP

Contents

Preface to the Third Edition, vii

Preface to the First Edition, viii

1. Basic Bacteriology: Structure, 1
2. Basic Bacteriology: Physiology, 9
3. Basic Bacteriology: Genetics, 14
4. Classification of Bacteria, 20
5. Staphylococci, 25
6. Streptococci and Enterococci, 30
7. Clostridia, 38
8. Other Gram-positive Bacteria, 42
9. Gram-negative Cocci, 48
10. Enterobacteriaceae, 51
11. Parvobacteria, 57
12. *Campylobacter, Helicobacter* and Vibrios, 63
13. *Pseudomonas* and other Aerobic Gram-negative Bacilli, 67
14. Gram-negative Anaerobic Bacteria, 71
15. *Treponema, Borrelia* and *Leptospira*, 73
16. Mycobacteria, 78
17. Chlamydiae, Rickettsiaceae, *Mycoplasma* and L-forms, 82
18. Basic Virology, 88
19. Major Virus Groups, 95
20. Basic Mycology and Classification of Fungi, 114
21. Parasitology: Protozoa, 122
22. Parasitology: Metazoa (Helminths), 137
23. Host–Parasite Relationships, 154

24 Immunology of Infection, 158
25 Diagnostic Laboratory Methods, 168
26 Antibacterial Agents, 177
27 Antifungal Agents, 194
28 Antiviral Agents, 197
29 Skin and Soft Tissue Infections, 202
30 Bone and Joint Infections, 212
31 Oral and Dental Infections, 216
32 Upper Respiratory Tract Infections, 220
33 Lower Respiratory Tract Infections, 226
34 Gastrointestinal Infections, 234
35 Liver and Biliary Tract Infections, 244
36 Urinary Tract Infections, 250
37 Genital Infections (Including Sexually Transmitted Diseases), 254
38 Infections of the Central Nervous System, 260
39 Septicaemia and Bacteraemia, 270
40 Infections of the Heart, 276
41 Perinatal and Congenital Infections, 281
42 Miscellaneous Viral Infections, 291
43 Infections in the Compromised Host, 297
44 Zoonoses, 301
45 Pyrexia of Unknown Origin, 308
46 Epidemiology and Prevention of Infection, 312
47 Immunization, 323
48 Sterilization and Disinfection, 329

Subject Index, 333

Organism Index, 341

Colour plates fall between pp. 184 and 185

Preface to the Third Edition

This book is written specifically for students, including medical, dental, science and pharmacology students studying medical microbiology. It will also serve as useful revision for doctors sitting the FRCS and MRCP examinations, and other health care professionals including MSOs and health science students.

It is divided into two main sections: organisms and infections; and covers all aspects of microbiology including bacteriology, virology, mycology and parasitology.

The text has been deliberately designed to highlight major points and important factors. Further information may be obtained from detailed reviews and textbooks.

We would like to thank Miss Helen Bailey for her ability to decipher early drafts and for typing the manuscript, also Miss Suzy Thompstone for help with typing sections of the book.

We are grateful to Mr Jim Brown for his assistance with the illustrations and to Glaxo-Wellcome for their educational grant.

Finally, we would like to thank the publishers Blackwell Science, in particular Andrew Robinson, for their cooperation and patience in the preparation of the manuscript.

Tom Elliott
Mark Hastings
Ulrich Desselberger

Preface to the First Edition

This small volume has been prepared in the hope that, in at least some of his lectures in bacteriology, the student will be saved the task of scribbling notes; the text is not meant to be exhaustive as will be obvious from the size of the volume in comparison with that of many other textbooks on the subject. An attempt has been made to highlight those features of bacterial species which are important in their identification and in the laboratory diagnosis of infection. Additionally, brief notes have been included on the epidemiology and prevention of certain infections.

CHAPTER 1

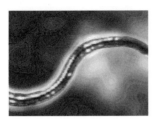

Basic Bacteriology: Structure

BACTERIAL STRUCTURE

Bacteria are approximately 0.1–10.0 μm long (Fig. 1.1). They can attain various shapes, including spheres (cocci), curves, spirals and rods (bacilli) (Fig. 1.2), which form a basis for classification. Bacterial structures (Fig. 1.3) are described below.

Cell envelope

Cytoplasmic membrane

Cytoplasmic membranes surround the cytoplasm of all bacteria cells and are composed of protein, phospholipid and a small amount of carbohydrate; they resemble the membrane surrounding mammalian cells. The phospholipids form a bilayer into which proteins are embedded, some spanning the membrane. The membrane carries out many functions, including the synthesis and export of cell-wall components, respiration, secretion of extracellular enzymes and toxins, and the uptake of nutrients by active transport mechanisms.

Mesosomes are intracellular membrane structures, formed by an invagination of the cytoplasmic membrane. They are more frequently seen in Gram-positive than Gram-negative bacteria. Mesosomes present at the septum of Gram-positive bacteria are involved in chromosomal separation; at other sites they may be associated with cellular metabolism.

Cell wall

Bacteria maintain their shape by a strong rigid outer cover, the cell wall (Fig. 1.3). Bacteria can be divided broadly into two main groups according to their Gram stain reaction which reflects the structure of their cell walls. Some bacteria stain Gram-positive (blue/black), whereas others are Gram-negative (red).

Gram-positive bacteria have a relatively thick cell wall, largely composed of peptidoglycan, a complex molecule consisting of repeating sugar sub-

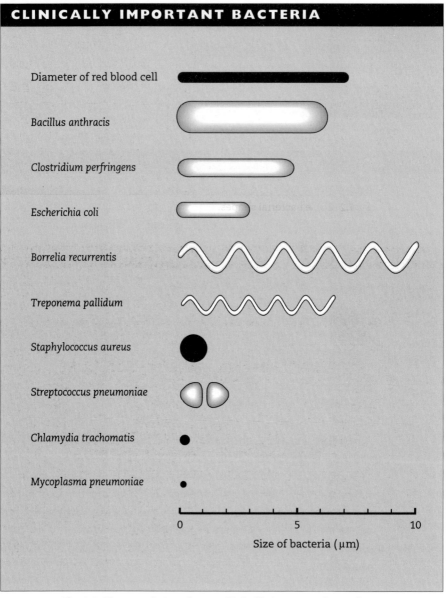

Fig. 1.1 Shape and size of some clinically important bacteria.

units cross-linked by peptide side-chains (Fig. 1.4a). Other cell-wall polymers, for example, teichoic acid, are also present.

Gram-negative bacteria have a thin peptidoglycan layer and an additional outer membrane which differs in structure from the cytoplasmic membrane (Fig. 1.4b). The outer membrane contains lipopolysaccharides, and

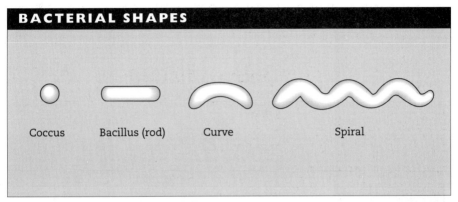

Fig. 1.2 Some bacterial shapes.

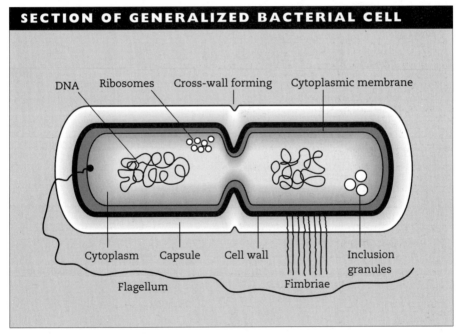

Fig. 1.3 A section of a typical bacterial cell.

in some species, lipoprotein and porins, which are proteins involved in transport of substances across the cell envelope. Lipopolysaccharides are a characteristic feature of Gram-negative bacteria and are also termed endotoxins. Endotoxins are released on cell lysis and have important biological activities involved in the pathogenesis of Gram-negative infections; they activate macrophages, clotting factors and complement, leading to disseminated intravascular coagulopathy and septic shock.

CELL WALL AND CYTOPLASMIC MEMBRANE

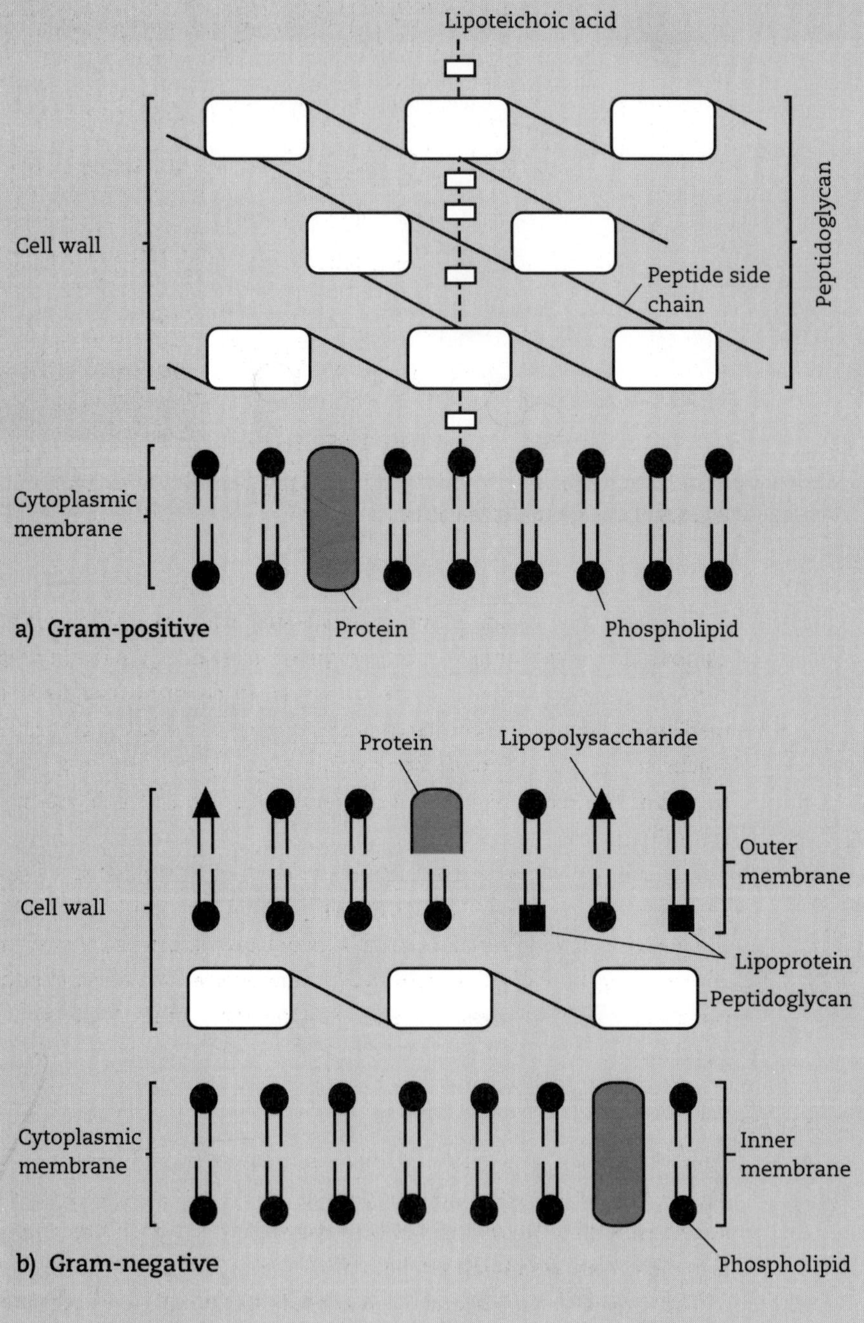

Fig. 1.4 Cell wall and cytoplasmic membrane of (a) Gram-positive and (b) Gram-negative bacteria. The Gram-positive cell wall has a thick peptidoglycan layer with membrane-bound lipoteichoic acid. The Gram-negative cell wall has lipopolysaccharides in an outer membrane, with a thin inner peptidoglycan layer.

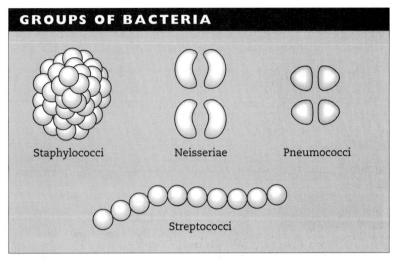

Fig. 1.5 Some groups of bacteria.

The cell wall is important in protecting bacteria against external osmotic pressure. Bacteria with damaged cell walls, e.g. following exposure to penicillin, often rupture. However, in an osmotically balanced medium, cell-wall-deficient bacteria may survive as spherical protoplasts. Under certain conditions some protoplasts can multiply and are referred to as L-forms. Some bacteria, e.g. mycoplasma, have no cell wall at any stage in their life cycle.

The cell wall is involved in bacterial division. After the nuclear material has replicated and separated, a cell wall (septum) forms at the equator of the parent cell. The septum grows in, produces a cross-wall and eventually the daughter cells may separate. In many species the cells can remain attached, forming groups, e.g. staphylococci form clusters and streptococci form long chains (Fig. 1.5).

Capsules

Some bacteria have capsules external to their cell walls (see Fig. 1.3). These capsules are firmly bound to the bacterial cell and have a compact structure with a clearly defined boundary. They are often composed of high molecular weight polysaccharides.

The capsules are important virulence determinants in both Gram-positive and Gram-negative bacteria. Their capsular antigens may be used to differentiate between isolates of the same bacteria, e.g. in the typing of *Streptococcus pneumoniae* for epidemiological purposes.

Bacterial slime

Extracellular slime layers are produced by some bacteria. They are more loosely bound to the cell surface than capsules and are water soluble. The slime is composed of complex polysaccharides. It is a virulence factor, e.g. facilitating the attachment of *Staphylococcus epidermidis* onto artificial surfaces, such as intravascular cannulae (Plate 1).

Flagella

Bacterial flagella are 3–14 μm long, thin (0.02 μm in diameter), spiral-shaped filaments consisting mainly of the protein, flagellin. They can be single (monotrichous) or multiple (peritrichous) (Fig. 1.6).

Flagella facilitate locomotion in all motile bacteria, except for the spirochaetes. They can be observed under the light microscope with special stains. However, the presence of flagella is usually detected for diagnostic purposes by observing motility in a bacterial suspension or by spreading growth on solid media.

Fimbriae

Fimbriae or pili are thin, hair-like appendages on the surface of many Gram-negative, and some Gram-positive, bacteria (see Fig. 1.3). They are approximately half the width of flagella, and are composed of proteins called pilins. In some bacteria they are distributed over the entire cell surface (>200/cell).

Fimbriae are virulence factors enabling bacteria to adhere to various mammalian cell surfaces, an important initial step in colonization of

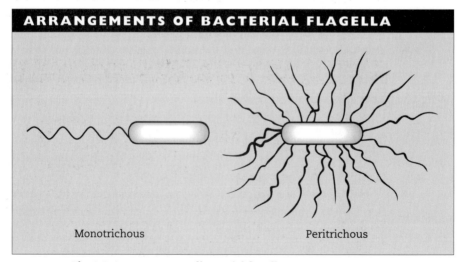

Fig. 1.6 Arrangements of bacterial flagella.

mucosal surfaces, e.g. *Neisseria gonorrhoeae* produce fimbriae that bind to specific receptors of cervical epithelial cells.

Specialized fimbriae are involved in genetic material transfer between bacteria, a process called conjugation.

Intracellular structures

Nuclear material

The bacterial nuclear material consists of a single circular molecule of double-stranded DNA, about 1mm long when unfolded. It is tightly packed within the bacterium and is *not* surrounded by a nuclear membrane as in mammalian cells. Smaller extra-chromosomal DNA molecules, called plasmids, that can replicate independently, may also be present. The chromosome usually codes for all the essential functions required by the cell; some plasmids control important phenotypic properties of pathogenic bacteria, including antibiotic resistance and toxin production.

Ribosomes

The cytoplasm has many ribosomes which contain RNA and proteins, and are involved in protein synthesis.

Inclusion granules

Various cellular inclusions which serve as energy and nutrient reserves may be present in the cytoplasm. The size of these inclusions can increase in a favourable environment and decrease when conditions are adverse, e.g. *Corynebacterium diphtheriae* may contain high energy phosphate reserves in inclusions termed volutin granules.

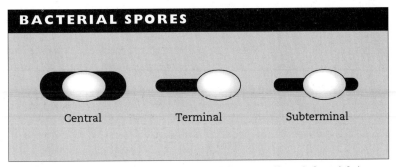

Fig. 1.7 Size, shape and position of bacterial spores: (from left to right) non-projecting, oval, central, e.g. *Bacillus anthracis*; projecting, spherical, terminal, e.g. *Clostridium tetani*; non-projecting, oval, subterminal, e.g. *C. perfringens*.

Endospores

Endospores (spores) are small, metabolically-dormant cells with a thick wall, formed intracellularly by members of the genera *Bacillus* and *Clostridium* (Plate 2). They are highly resistant to adverse environmental conditions and may survive desiccation, disinfectants or boiling water for several hours.

Spores are formed in response to limitations of nutrients by a complex process (sporulation) involving at least seven stages. When fully formed, they appear as oval or round cells within the vegetative cell. The location is variable, but is constant in any one bacterial species (Fig. 1.7). Spores can remain dormant for long periods. However, they are able to germinate relatively rapidly in response to certain conditions such as the presence of specific sugars.

Spores have an important role in the epidemiology of certain human diseases such as anthrax, tetanus and gas gangrene.

The irradication of spores is of particular importance in some processes, e.g. in the sterilization of instruments prior to surgery and in food-canning for the removal of *Clostridium botulinum* to prevent botulism.

CHAPTER 2

Basic Bacteriology: Physiology

BACTERIAL GROWTH

Most bacteria will grow on artificial culture media. However, some bacteria, e.g. *Mycobacterium leprae* (leprosy) and *Treponema pallidum* (syphilis), cannot yet be grown *in vitro*; other bacteria, e.g. chlamydiae and rickettsiae, only replicate within host cells and are grown in tissue-culture.

Under suitable conditions (nutrients, temperature and atmosphere) a bacterial cell will increase in size and then divide into two identical cells. These two cells are able to grow and divide at the same rate as the parent cell, providing conditions remain stable. This results in an exponential or logarithmic growth rate (Fig. 2.1). The time required for the number of bacteria in a culture to double is called the generation time; e.g. *Escherichia coli* has a generation time of about 30 min under optimal conditions.

Requirements for bacterial growth

Most bacteria of medical importance require carbon, nitrogen, water, inorganic salts and a source of energy for growth. They have various gaseous, temperature and pH requirements, and can utilize a range of carbon, nitrogen and energy sources. Some bacteria also require special growth factors, including amino acids and vitamins.

Growth requirements are important in selecting the various culture media required in diagnostic microbiology and in understanding the tests for identifying bacteria.

CARBON AND NITROGEN SOURCES

Bacteria are classified into two main groups according to the type of compounds they can utilize as a carbon source.

1 *Autotrophs* utilize inorganic carbon from carbon dioxide and nitrogen from ammonia, nitrites and nitrates; they are of minor medical importance.

2 *Heterotrophs* require organic compounds as their major source of carbon and energy; they include most bacteria of medical importance.

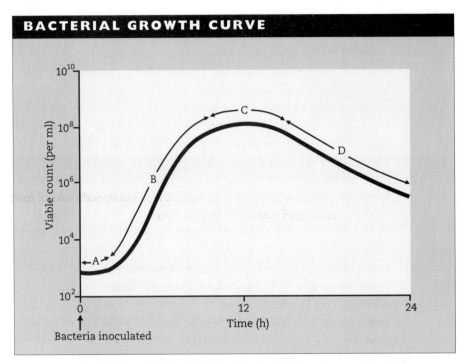

Fig. 2.1 Bacterial growth curve showing the four phases: A, lag; B, log or exponential; C, stationary; D, decline (death).

ATMOSPHERIC CONDITIONS
Carbon dioxide (CO_2)

Bacteria require CO_2 for growth; adequate amounts are present in air or are produced during metabolism by the organisms themselves. A few bacteria require additional CO_2 for growth, e.g. *Neisseria meningitidis*.

Oxygen (O_2)

Bacteria may be classified into four groups according to their O_2 requirements:

- *obligate (strict) aerobe*: grow only in the presence of oxygen, e.g. *Pseudomonas aeruginosa*;
- *microaerophilic bacteria*: grow better in low oxygen concentrations, e.g. *Campylobacter jejuni*;
- *obligate (strict) anaerobe*: grow only in the absence of free oxygen, e.g. *Clostridium tetani*;
- *facultative anaerobes*: grow in the presence or absence of oxygen, e.g. *E. coli*.

TEMPERATURE

Bacteria can be classified according to the optimal temperature for growth:
- psychrophiles: low temperatures (optima <20°C; some below 0°C);
- mesophiles: 20–45°C;
- thermophiles: >45°C.

Nearly all pathogenic bacteria grow best at 37°C. The optimum temperature for growth is occasionally higher, for example, for *C. jejuni*, it is 42°C. The ability of some bacteria to grow at low temperatures (0–4°C) is important in food microbiology; *Listeria monocytogenes*, a cause of food poisoning, will grow slowly at 4°C and has resulted in outbreaks of food poisoning associated with cook–chill products.

pH

Most pathogenic bacteria grow best in slightly alkaline pH (pH 7.2–7.6). A few exceptions exist; *Lactobacillus acidophilus*, present in the vagina of postpubescent females, prefers an acid medium (pH 4.0). It produces lactic acid which keeps the vaginal secretions acid, thus preventing many pathogenic bacteria from establishing infection. *Vibrio cholerae*, the cause of cholera, prefers an alkaline environment (pH 8.5).

Growth in liquid media

When bacteria are added (inoculated) into a liquid growth medium, subsequent multiplication can be followed by determining the total number of viable organisms (viable counts) at various time intervals. The growth curve produced normally has four phases (Fig. 2.1).

1 Lag Phase (A): the interval between inoculation of a fresh growth medium with bacteria and the commencement of growth.

2 Log Phase (B): the phase of exponential growth; the growth medium becomes visibly turbid at approximately 10^4 cells/ml.

3 Stationary Phase (C): the growth rate slows as nutrients become exhausted, waste products accumulate, and the rate of cell division equals the rate of death; the total viable count remains relatively constant.

4 Decline Phase (D): the rate of bacterial division is slower than the rate of death, resulting in the decline in total viable count.

Note that the production of waste products by bacteria, particularly CO_2, and the uptake of O_2, have been utilized in the development of modern instruments to detect bacterial growth in blood cultures obtained from patients with septicaemia.

Growth on solid media

Liquid media can be solidified with agar which is extracted from algae. A temperature of 100°C is used to melt agar, which then remains liquid until the temperature falls to approximately 40°C, when it produces a transparent gel. Solid media are normally set in Petri dishes ('agar plates'). Most bacteria grow on solid media to produce colonies. Each colony comprises thousands of bacterial cells which emanated from a single cell. The morphology of the colony assists in bacterial identification.

Growth on laboratory media

To culture bacteria *in vitro*, the microbiologist has to take into account the physiological requirements. Various types of liquid and solid media have been developed for the diagnostic laboratory.

SIMPLE MEDIA

Many bacteria will grow in simple media, e.g. nutrient broth/nutrient agar which contains 'peptone' (polypeptides and amino acids from the enzymatic digestion of meat) and 'meat extract' (water-soluble components of meat containing mineral salts and vitamins).

ENRICHED MEDIA

These contain additional nutrients for the isolation of fastidious bacteria which require special conditions, e.g. blood agar; chocolate agar (heated to lyse erythrocytes and release additional nutrients).

SELECTIVE MEDIA

These are designed to facilitate growth of some bacteria, whilst suppressing the growth of others and include: *MacConkey agar*, which contains bile salts and allows the growth of bile-tolerant bacteria only; *alkaline peptone broth*, which allows selective growth of vibrios which prefer alkaline conditions; and antibiotics, which are frequently added to media to allow only certain bacteria to grow whilst suppressing or killing others.

INDICATOR MEDIA

These are designed to aid the detection and recognition of particular pathogens. They are often based on sugar fermentation reactions which result in production of acid and the subsequent colour change of a pH indicator, e.g. MacConkey agar contains lactose and a pH indicator (neutral red); lactose-fermenting bacteria (e.g. *E. coli*) produce acid and form pink colonies, whilst non-lactose fermenting bacteria (e.g. salmonella) do not produce acid and form pale yellow colonies. This property facilitates the recognition of possible salmonella colonies amongst normal bowel flora.

CHAPTER 3

Basic Bacteriology: Genetics

REPLICATION OF BACTERIAL DNA

The normal mechanism of reproduction in bacteria is by binary fission. Genetic information is carried on double-stranded DNA, which, although arranged in a circular manner, is folded multiple times to fit inside the cell. An enzyme, DNA polymerase, is important in this folding process; the quinolone antibiotics inhibit this enzyme.

The structure and replication of bacterial DNA is similar to that of eukaryotic cells. Bacterial DNA consists of two complementary strands and during cell division, these two strands separate and each acts as a template allowing the formation of two new double-stranded molecules.

PROTEIN SYNTHESIS

DNA is transcribed into RNA, which is in turn translated into new proteins via messenger RNA (mRNA) and ribosomes (Fig. 3.1). The molecular basis of transcription and translation is important in the understanding of the action of some antibiotics, e.g. gentamicin, rifampicin and erythromycin, which inhibit protein synthesis.

The main steps of protein synthesis are:

1 *transcription*: mRNA is transcribed from the chromosomal DNA strand;
2 *translation*: codons (the nucleotide triplets) in mRNA specify the amino acid to be inserted in the forming polypeptide during translation; are translated by binding of transfer RNA (tRNA) with the formation of a protein directed by a specific mRNA molecule displaying appropriate corresponding nucleotides for each codon; the tRNA carries a matching amino acid:
- the ribosome binds to the mRNA;
- an enzyme peptidyl transferase facilitates the transfer of each new amino acid to the growing peptide chain;
- the mRNA is sequentially translated by this mechanism until a 'nonsense' codon is encountered which results in termination of the peptide chain.

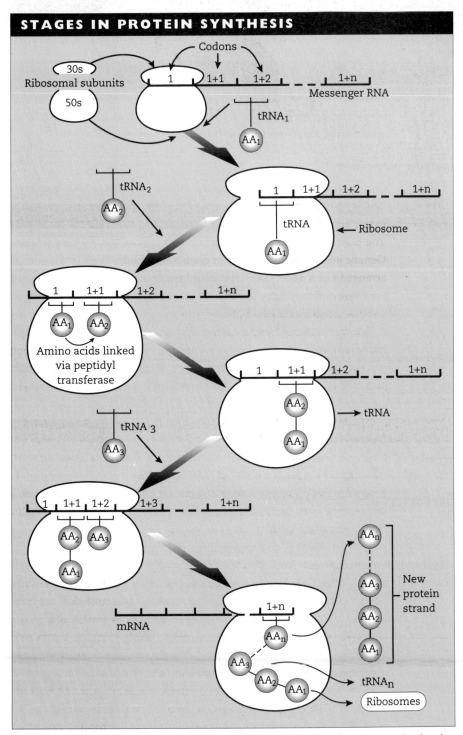

Fig. 3.1 The major stages in protein synthesis involving ribosomal subunits (30s and 50s), messenger RNA (mRNA), transfer RNA (tRNA) and amino acid blocks (AA$_n$).

GENOTYPIC VARIATION

The ability of bacteria to alter genetic information is fundamental to their survival in a changing environment. Such variation in the genome can occur in two ways, mutation and recombination (genetic transfer).

MUTATION

During replication of DNA, copying errors called mutations may occur leading to changes in the sequence of nucleic acids. Mutations can arise spontaneously or by exposing bacteria to radiation or mutagenic chemicals.

Many mutations result in a change in only a single nucleotide, with no detectable alteration in the end product, the transcribed protein. These are called point mutations.

More major changes, which lead to significant alterations in the organism, are often detrimental and the mutant organism may not survive. However, under certain circumstances, such alterations can result in a mutant cell with a significant advantage, allowing it to outgrow other daughter cells; for example, antibiotic resistance mutants may be selected out when that particular antibiotic is present in the environment. The principles of mutation and environmental selection are universal to biology.

RECOMBINATION (GENETIC TRANSFER)

There are three basic mechanisms whereby new genetic information can transfer from one bacterium to another: transformation, conjugation and transduction (Fig. 3.2).

Transformation

Some bacteria are able to take up soluble DNA fragments derived from other, normally closely related, species directly across their cell wall. This process of transformation was first described in *Streptococcus pneumoniae*.

Transduction

Viruses that infect bacteria are known as bacteriophages. During replication in bacteria, bacteriophages may incorporate some host DNA into the capsid. When the virus is released and infects a new bacterial cell, the bacterial DNA from the donor cell may be integrated into the chromosome of the recipient. Bacteriophages infect only a narrow range of bacteria and thus this form of DNA recombination can only occur between closely related bacterial strains.

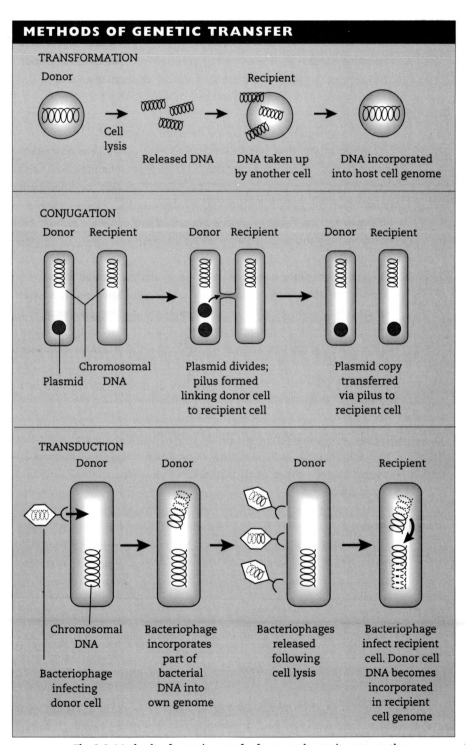

Fig. 3.2 Methods of genetic transfer from one bacterium to another.

A bacteriophage can also incorporate its own viral DNA into the bacterial chromosome; occasionally, this can result in the bacteria synthesizing new proteins; diphtheria toxin and the erythrogenic toxin of group A β-haemolytic streptococci are both encoded on phage DNA.

Conjugation

Extrachromosomal segments of DNA, called *plasmids*, are present in some bacteria and are distinct from the chromosome. Plasmids replicate independently of the chromosome and, therefore, must encode certain essential genes necessary for DNA replication. Plasmids vary considerably in size, from large (80–120 kilobases (kb)) to small (2–10 kb).

Genes carried on plasmids may control one or more phenotypic characteristics of the bacterial cell, including antibiotic resistance and synthesis of bacterial toxins.

Certain plasmids, known as *conjugative plasmids*, are involved in the transfer of DNA between bacteria. Conjugative plasmids contain genes which control the formation of pili. These allow the bacterial cell to attach to a second cell via a cytoplasmic bridge. The conjugative plasmid divides and a copy is transferred across the cytoplasmic bridge into the recipient cell. Other non-conjugative plasmids can also be transferred in this way, provided the host bacteria contains a conjugative plasmid to initiate the process.

The transfer of genetic information via plasmids can occur rapidly and between bacteria that are not closely related. Plasmid transfer of antibiotic resistant genes is the most important mechanism by which resistance to antibiotics can spread. Some plasmids carry as many as six different antibiotic resistant genes.

Moveable genetic elements known as *transposons* ('jumping genes') are small genetic elements which can transfer between plasmids and chromosomal DNA within cells. At each end of the transposon are specific base sequences known as insertion sequences which allow the transposon DNA to be inserted into existing DNA strands. Transposons allow genetic information to be transferred rapidly between plasmids and chromosomal DNA, facilitating the dissemination of genetic information amongst a bacterial population.

GENETIC ENGINEERING

The various vectors that allow the transfer of DNA between bacteria form the basis of genetic engineering. Genes coding for important products can be inserted into plasmids or bacteriophages and used to transfect other organisms of the same or different species allowing the synthesis of large amounts of the defined product. These techniques have been utilized in the field of vaccine research and production, and for studying bacterial pathogenicity and resistance to antimicrobial agents.

CHAPTER 4

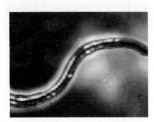

Classification of Bacteria

BACTERIAL TAXONOMY AND NOMENCLATURE

The classification of microorganisms is essential for the understanding of clinical microbiology. Bacteria are designated by a binomial system, with the genus name (capital letter) followed by the species name (without capital letter), for example, *Escherichia coli* and *Staphylococcus aureus*. Names are often abbreviated, for example, *E. coli* and *S. aureus*.

Many nomenclature problems exist with this system, which can lead to confusion; for example, 'bacillus' refer to any rod-shaped bacteria, whereas the genus *Bacillus* includes only the aerobic spore-bearing rods. Other complications include the use of alternative terminology. *Streptococcus pneumoniae*, is referred to as the pneumococcus and *Neisseria meningitidis* as the meningococcus. Occasionally, collective terms are used, for example, the term 'coliform' may indicate *E. coli* or a closely related bacterium; the term 'coagulase-negative staphylococci' means staphylococci which are not *S. aureus*. In this text, conventional terminology is used and, where appropriate, common alternatives are indicated.

BACTERIAL CLASSIFICATION

Medically important bacteria can be subdivided into five main groups according to their morphology and staining reactions. The basic shape of bacteria include cocci, bacilli, spiral and pleomorphic forms. Each of these morphological forms is further subdivided by their staining reactions, predominantly the Gram and acid-fast stains (Table 4.1). The more important pathogenic bacteria are divided primarily into Gram-positive or Gram-negative organisms. Other characteristics, including the ability to grow in the presence (aerobic) or absence of oxygen (anaerobic) and spore formation, are used further to divide the groups. Subdivision of these groups into genera is made on the basis of various factors, including culture properties (e.g. conditions required for growth and colonial morphology), antigenic properties and biochemical reactions. The medically important

PATHOGENIC BACTERIA

I   Gram-positive cocci, bacilli and branching bacteria
II Gram-negative cocci, bacilli and comma-shaped bacteria
III Spiral-shaped bacteria
IV Acid-fast bacteria
V   Cell-wall-deficient bacteria

Table 4.1 Main groups of pathogenic bacteria.

GRAM-POSITIVE BACTERIA

Grouping	Aerobic/anaerobic growth	Genus	Examples of clinically important species
Gram-positive cocci			
Clusters	Both	Staphylococcus	S. aureus
			S. epidermidis
			S. saprophyticus
Chains/pairs	Both	Streptococcus	S. pneumoniae
			S. pyogenes
			E. faecalis
Squares	Both	Micrococcus	
Chains	Anaerobic	Peptococcus and Peptostreptococcus	
Gram-positive bacilli			
Sporing	Aerobic	Bacillus	B. anthracis
			B. cereus
Non-sporing	Both	Corynebacterium	C. diphtheriae
	Aerobic or microaerophilic	Listeria	L. monocytogenes
	Anaerobic or microaerophilic	Lactobacillus	
Sporing	Anaerobic	Clostridium	C. difficile
			C. botulinum
			C. perfringens
			C. tetani
Non-sporing	Anaerobic	Propionibacterium	P. acnes
Branching	Anaerobic	Actinomyces	A. israeli
	Aerobic	Nocardia	N. asteroides

Table 4.2 Classification of major groups of bacterial pathogens.

GRAM-NEGATIVE BACTERIA

Shape	Aerobic/ anaerobic growth	Major grouping	Genus	Examples of clinically important species
Cocci	Aerobic		Neisseria	N. gonorrhoeae N. meningitidis
Cocci	Anaerobic		Veillonella	
Bacilli		Enterobacteriaceae	Enterobacter Escherichia Klebsiella Proteus Salmonella Serratia Shigella Yersinia	E. chloaceae E. coli K. pneumoniae P. mirabilis S. typhimurium S. marcescens S. sonnei Y. enterocolitica
Bacilli	Aerobic		Pseudomonas	P. aeruginosa
Comma-shaped	Both	Vibrios	Vibrio Campylobacter Helicobacter	V. parahaemolyticus V. cholerae C. jejuni H. pylori
Bacilli	Varies with genus	Parvobacteria	Bordetella Brucella Haemophilus Pasteurella	B. pertussis B. abortus H. influenzae H. parainfluenza P. multocida
Bacilli	Aerobic		Legionella	L. pneumophila
Bacilli	Anaerobic		Bacteroides Fusobacterium	B. fragilis

Table 4.3 Classification of Gram-negative bacterial pathogens.

genera based on this classification are shown in Table 4.2 (Gram-positives) and Table 4.3 (Gram-negatives).

OTHER BACTERIAL GROUPS

Spiral bacteria

These are relatively slender spiral-shaped filaments, which are classified into three clinically important genera.

1 *Borrelia*: these are relatively large, motile spirochaetes and include *Borrelia vincenti* and *Leptotrichia buccalis*, which cause Vincent's angina, and *B. recurrentis*, which causes relapsing fever.

2 *Treponema*: these are thinner and more tightly spiralled than *Borrelia*. Examples include *Treponema pallidum* (causes syphilis) and *T. pertenue* (causes Yaws).

3 *Leptospira*: these are finer and are even more tightly coiled than the Treponemes. They are classified within the single species of *Leptospira interrogans*, which is divided serologically into two complexes. There are over 130 serotypes in the interrogans complex, many of which are pathogenic, including *L. icterohaemorrhagiae* (causes Weil's disease) and *L. canicola* (causes lymphocytic meningitis).

Acid-fast bacilli

These include the genus *Mycobacterium*. They are identified by their acid-fast staining, which reflects their ability to resist, after being stained with hot carbol fuchsin, decolorization with acid. Mycobacteria are generally difficult to stain by Gram's method. They can be simply divided into the following main groups:

1 Tubercle bacilli: *Mycobacterium tuberculosis* and *M. bovis*;

2 Leprosy bacillus: *M. leprae*;

3 *Atypical mycobacteria*. Some tuberculosis-like illnesses in man are due to other species of mycobacteria. They can grow at 27°C, 42°C or 45°C; some produce pigment when growing in light and are called photochromogens, whereas others produce pigment in light or darkness and are referred to as scotochromogens. Other mycobacteria can be rapid growers. All these species of mycobacteria are commonly referred to as the atypical mycobacteria; examples include *M. kansasii*, photochromogenic; *M. avium-intracellulare*, non-pigmented and *M. chelonei*, fast growing.

Cell-wall-deficient bacteria

Some bacteria do not form cell walls and are called mycoplasmas. Pathogenic species include *Mycoplasma pneumoniae* and *Ureaplasma urealyticum*. It is important to distinguish mycoplasmas from other cell-wall-deficient forms of bacteria, which can be defined as either L-forms or protoplasts.

• *L-forms* are cell-wall-deficient forms of bacteria, which are produced by removal of a bacteria's cell wall, for example with cell-wall-acting antibiotics. L-forms are able to multiply and their colonial morphology is similar to the 'fried egg' appearance of the mycoplasmas.

- *Protoplasts* are bacteria which have also had their cell walls removed. They are metabolically active and can grow, but are unable to multiply. They only survive in an osmotically stabilized medium. *Spheroplasts* are protoplasts but with some cell wall remaining.

CHAPTER 5

Staphylococci

There are at least 20 species of staphylococci, but only three are clinically important: *Staphylococcus aureus*, *S. epidermidis* (*S. albus*) and *S. saprophyticus*. Their principal characteristics are shown in Table 5.1.

Definition. Gram-positive cocci; usually arranged in clusters (Plate 3); non-motile; non-sporing; usually non-capsulate; grow over a wide temperature range (10–42°C) with an optimum of 37°C; aerobic and facultatively anaerobic; grow on simple media.

Classification. *S. aureus* is distinguished from other staphylococci on the basis of the following factors.
1 *Colonial morphology*: *S. aureus* colonies are grey to golden yellow (Plate 4); *S. epidermidis* and *S. saprophyticus* colonies are white. *S. aureus* produces haemolysins, resulting in haemolysis on blood agar.
2 *Coagulase test*: *S. aureus* possesses the enzyme coagulase which acts on plasma to form a clot. Other staphylococci (e.g. *S. epidermidis*) do not possess this enzyme and are often termed collectively, coagulase-negative staphylococci. There are two methods to demonstrate the presence of coagulase:
    (a) tube coagulase test: diluted plasma is mixed with a suspension of the bacteria; following incubation, clot formation indicates *S. aureus*;
    (b) slide coagulase test: a more rapid and simple method in which a drop of plasma is added to a suspension of staphylococci on a glass slide; visible clumping indicates the presence of coagulase.
3 *Deoxyribonuclease (DNAase) production*: *S. aureus* possesses an enzyme DNAase; other staphylococci rarely possess this enzyme.
4 *Protein A detection*: *S. aureus* possesses a cell-wall antigen, protein A; antibodies to protein A agglutinate *S. aureus* but not other staphylococci.

S. aureus

Epidemiology. *S. aureus* is a relatively common commensal of man: nasal carriage occurs in 30–50% of healthy adults, faecal carriage in about 20%

and skin carriage in 5–10% particularly axilla and perineum. *S. aureus* is spread via droplets and skin scales which contaminate clothing, bedlinen and other environmental sources.

Morphology and identification. On microscopy *S. aureus* is seen as typical Gram-positive cocci in clusters. It is coagulase- and DNAase-positive. Other biochemical tests can be performed.

Pathogenicity. Staphylococci produce disease because of their ability to; spread in tissues and form abscesses, produce extracellular enzymes or exotoxins (Table 5.2) and combat host defences.

Exotoxins and enzymes
- *Coagulase:* *S. aureus* produces coagulase, an enzyme that coagulates

CHARACTERISTICS OF STAPHYLOCOCCI

Characteristic	S. aureus	S. epidermidis	S. saprophyticus
Coagulase	+	−	−
Deoxyribonuclease	+	−	−
Colonial appearance; smooth, shiny, opaque (majority)	Golden-yellow	White	White
Locations as commensal	Nose, Mucosal surfaces, Faeces, Skin	Skin, Mucosal surfaces	Periurethral, Faeces
Common infections	Skin (boils, impetigo, furuncles, wound infections) Abscesses Osteomyelitis Septic arthritis Septicaemia Infective endocarditis	Infections of prosthetic devices, e.g. artificial heart valves, intravenous catheters, CSF shunts	Urinary tract infections in sexually active young women

+, present; −, absent; CSF, cerebrospinal fluid.

Table 5.1 Characteristics of staphylococci.

TOXIN PRODUCTION	
Toxin*	Effect
Enterotoxins	Released into foods and results in food poisoning
Exfoliative or epidermolytic	Scalded skin syndrome, producing peeling of layers of epidermis
Haemolysins	Lyse erythrocytes
Leucocidins	Lyse leucocytes and macrophages
Toxic shock syndrome	Vascular collapse, rash with desquamation

* Toxin production varies between strains of S. aureus.

Table 5.2 Toxin production associated with S. aureus.

plasma. Coagulase results in fibrin deposition which interferes with phagocytosis and increases the ability of the organism to invade tissue.
- *Haemolysins and leucocidin:* several exotoxins are produced by *S. aureus*; α-toxin (haemolysin), lyses erythrocytes and damages platelets; β-toxin degrades sphingomyelin and is toxic for many types of cell, including erythrocytes; leucocidin lyses white blood cells.
- *Other enzymes:* S. *aureus* may also produce staphylokinase (which can result in fibrinolysis), hyaluronidase (dissolves hyaline), proteinases (degrade proteins), and lipases (solubilize lipids).
- *Enterotoxins:* there are six soluble enterotoxins which are produced by nearly half of all *S. aureus* strains. They are heat stable (resistant at 100°C for 30 min), unaffected by gastrointestinal enzymes and are a cause of food poisoning, principally associated with vomiting.
- *Exfoliative toxin:* some strains produce a toxin that can result in generalized desquamation of the skin (staphylococcal scalded skin syndrome).
- *Toxic shock syndrome toxin:* this is associated with shock, desquamation of skin, and is usually related to an underlying *S. aureus* infection.

Cell envelope

Some *S. aureus* strains possess capsules which interfere with opsonization and phagocytosis. *S. aureus* also possesses a cell-wall protein (protein A) which binds the F_c component of antibody preventing complement activation.

Laboratory diagnosis. Laboratory diagnosis is by direct isolation from the infected site or from blood cultures and by detection of serum antibodies to staphylolysin and DNAase.

S. aureus strains can be typed ('fingerprinted') by the use of bacteriophages (phage typing, p. 93).

Antibiotic therapy. Antimicrobial agents such as flucloxacillin (≡ methicillin) or erythromycin with fusidic acid remain the first-line treatment for *S. aureus* infections; the emergence of methicillin-resistant *S. aureus* (MRSA) has required the use of glycopeptide antibiotics such as vancomycin. MRSA can cause sepsis, ranging from wound infections, urinary tract infections and septicaemia. Epidemic strains of MRSA (EMRSA) have also been recognized. Prevention of spread is therefore important.

Associated infections
- Skin: boils, impetigo, furuncles, wound infections, staphylococcal scalded skin syndrome.
- Respiratory: pneumonia, lung abscesses, exacerbations of chronic lung disease.
- Skeletal: most common cause of osteomyelitis and septic arthritis.
- Invasive: septicaemia (including infective endocarditis), deep abscesses (brain, liver, spleen), toxic shock syndrome.
- Gastrointestinal: toxin-mediated food poisoning.

S. epidermidis

- *S. epidermidis* is coagulase- and DNAase-negative and is present in large numbers on the human skin and mucous membranes.
- *S. epidermidis* is a cause of bacterial endocarditis, particularly involving prosthetic valves and drug addicts. It is also associated with infections of implanted plastic devices such as cerebrospinal shunts, hip prostheses, central venous and peritoneal dialysis catheters.
- The organism attaches firmly onto artificial surfaces. Some strains produce a glycocalyx ('slime layer') which appears to facilitate adhesion and protect the organism from antibiotics and host defences. The increased use of implanted devices, particularly central venous catheters, has resulted in *S. epidermidis* becoming one of the most frequently isolated organisms from blood cultures. *S. epidermidis* occasionally causes urinary tract infections, particularly in catheterized patients. When isolated from hospitalized patients *S. epidermidis* is often resistant to antibiotics such as flucloxacillin and erythromycin, necessitating the use of glycopeptide antibiotics (e.g. vancomycin).

S. saprophyticus

S. saprophyticus is coagulase- and DNAase-negative and is associated with urinary tract infections in sexually active young women, occasionally resulting in severe cystitis with haematuria.

CHAPTER 6

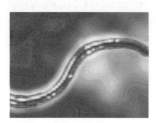

Streptococci and Enterococci

Streptococci are Gram-positive cocci, which are arranged in pairs or chains (Plate 5). They are widely distributed and are commensals of the human upper respiratory, gastrointestinal and female genital tracts. Certain species are pathogenic. Table 6.1 lists the major infections caused by streptococci.

Note: Streptococcus faecalis (faecal streptococci; enterococci) has now been placed in a separate genus, *Enterococcus*, and is described at the end of this chapter together with the anaerobic streptococci (*Peptococcus* and *Peptostreptococcus*).

Definition. Gram-positive cocci; non-motile; non-sporing; occasionally capsulate; optimum growth at 37°C; facultative aerobes; often require enriched media; many species show characteristic haemolysis.

Classification. The classification of streptococci is based on the following factors.

1 *Haemolysis:* streptococci are alpha (α), beta (β) or non-haemolytic.
   (a) α-haemolysis is the formation around colonies on blood agar of a greenish zone in which erythrocytes are disrupted and the haemoglobin converted to a green pigment (Plate 6). Examples of α-haemolytic streptococci include the viridans streptococci and *S. pneumoniae*.
   (b) β-haemolysis is the formation around colonies on blood agar of a clear zone in which erythrocytes have been completely lysed (Plate 7). The degree of haemolysis can vary between strains. β-haemolytic streptococci are classified further according to cell-wall antigens.
   (c) Non-haemolytic streptococci are unable to haemolyse erythrocytes and have no affect on blood agar.

2 *Cell envelope antigens:* in the 1930s, Lancefield classified β-haemolytic streptococci alphabetically according to the possession of carbohydrate antigens in the cell envelope. Specific antibodies to each of the streptococcal group antigens were used in agglutination tests. This still forms the basis of modern laboratory techniques for the routine grouping of β-haemolytic streptococci. The medically important β-haemolytic

INFECTIONS CAUSED BY STREPTOCOCCI	
Streptococci	Major clinical syndromes
β-Haemolytic	
Group A	Pharyngitis
	Cellulitis, erysipelas, necrotizing fasciitis
	Septicaemia
	Rheumatic fever
	Acute glomerulonephritis
	Scarlet fever
Group B	Neonatal meningitis and septicaemia
Group C	Pharyngitis
	Cellulitis
Group G	Cellulitis
α-Haemolytic	
Viridans streptococci	Dental caries
	Endocarditis
S. pneumoniae	Otitis media/sinusitis
	Meningitis
	Pneumonia
Others	
S. milleri	Deep abscesses
Anaerobic streptococci	Abscesses

Table 6.1 Infections caused by streptococci.

streptococci groups are A, B, C, F and G. Group D streptococci (*S. faecalis*) have been reclassified in the genus *Enterococcus*.

3 *Biochemical reactions*: some streptococci are difficult to classify by haemolytic and antigenic characteristics; biochemical tests are used in their identification.

β-HAEMOLYTIC STREPTOCOCCI

Group A β-haemolytic streptococci (S. pyogenes)

Epidemiology. Group A β-haemolytic streptococci are upper respiratory tract commensals of 3–5% of adults and up to 10% of children. Transmission is principally via droplet spread.

Morphology and identification. They are facultative aerobic organisms which grow best on enriched media containing blood; typically they show a large zone of clear β-haemolysis around colonies. The identity is confirmed by Lancefield grouping and biochemical reactions.

Pathogenicity. Group A β-haemolytic streptococci demonstrate a range of virulence factors.
- *Lipoteichoic acid:* this is an adherence factor which allows adherence to pharyngeal epithelial cells.
- *M-proteins:* these are surface proteins which act as anti-phagocytic factors. Different strains possess different M-proteins and this may be used for epidemiological typing. Antibodies to M-proteins are protective against infection.
- *Haemolysins and leucocidins:* these are exotoxins which lyse erythrocytes and leukocytes (streptolysins). They are responsible for β-haemolysis seen on blood agar.
- *Exotoxins:* these include streptokinase (which facilitates the breakdown of fibrin and allows spread of streptococci through tissues), hyaluronidase (which breaks down hyaluronic acid in connective tissue facilitating spread), and deoxyribonucleases.
- *Erythrogenic toxin:* this is responsible for the rash of scarlet fever. Production is dependent upon the presence of a bacteriophage which codes for the toxin.

Associated infections
- *Upper respiratory tract:* pharyngitis, tonsillitis and otitis media.
- *Skin:* cellulitis, necrotizing fasciitis, impetigo, erysipelas, wound infection and scarlet fever.
- *Invasive:* septicaemia and puerperal sepsis.

Post-infection complications. Rheumatic fever and acute glomerulonephritis are post-streptococcal, immune-mediated complications, which develop up to 3 weeks after the primary infection. In rheumatic fever a cross-reaction between Group A streptococcal cell wall antigen and cardiac muscle occurs. In acute post-streptococcal glomerulonephritis, immune complexes of streptococcal antigen and antibody are deposited in the glomeruli; the resulting complement activation and inflammatory response result in glomerulonephritis.

Laboratory diagnosis
- Diagnosis is by isolation of the organism from infected sites (e.g. throat, skin) or blood cultures. Direct detection of antigen in throat swabs, by a monoclonal antibody–agglutination reaction has recently been developed but is not in widespread use.
- Retrospective diagnosis is by detection of serum antibodies to streptolysin O (ASO). This is important in the diagnosis of rheumatic fever and

acute glomerulonephritis as the organism is often no longer present at the time of presentation.
- M-proteins of Group A β-haemolytic streptococci can be used for epidemiological typing.

Antibiotic therapy. Penicillin (erythromycin for penicillin-allergic patients) remains the drug of choice for treatment of infection by Group A β-haemolytic streptococci.

Group B β-haemolytic streptococci

Epidemiology. Group B β-haemolytic streptococci form part of the normal perineal flora in about 30–40% of individuals.

Morphology and identification. They grow readily on blood agar and are identified by Lancefield grouping.

Pathogenicity. The virulence factors are less well defined than for Group A β-haemolytic streptococci. Type-specific antigens (carbohydrate and protein) appear to be important in virulence (analogous to M-proteins of the Group A streptococci).

Associated infections
- *Neonates*: Group A streptococci are an important cause of septicaemia and meningitis.
- *Adults*: they are an occasional cause of urinary tract and genital infections, post-partum sepsis, septicaemia (particularly in the elderly) and endocarditis.

Laboratory diagnosis. This is by isolation from the infected site, e.g. blood, cerebrospinal fluid (CSF). Direct detection of the antigen from CSF or urine by a monoclonal antibody-agglutination reaction may be performed.

Antibiotic therapy. Penicillin or erythromycin are the drugs of choice for treatment of Group B β-haemolytic streptococcal infection.

Other β-haemolytic streptococci

Group C streptococci contain a number of different species, some of which are important animal pathogens. Infections in man occur and are similar to those caused by Group A streptococci.

Group G streptococci occasionally cause skin infections in man.

α-HAEMOLYTIC STREPTOCOCCI

The α-haemolytic streptococci are classified into *S. pneumoniae* (pneumococcus) and the viridans streptococci.

Occasional isolates of α-haemolytic streptococci may possess Lancefield carbohydrate antigens.

S. pneumoniae

Epidemiology. *S. pneumoniae* is a normal inhabitant of the upper respiratory tract.

Morphology and identification. Gram-positive cocci, often in pairs with a capsule. They are facultative aerobic organisms with enhanced growth in the presence of 10% carbon dioxide (CO_2). Colonies are typically disc-shaped with raised edges ('draughtsmen') (Plate 8) surrounded by a zone of α-haemolysis. Pneumococci are identified by their sensitivity to optochin.

Pathogenicity. Pneumococci possess a polysaccharide capsule of which there are over 80 antigenic types. Type-specific antibody to capsular antigen is required by the host for efficient opsonization and phagocytosis of the organism. Of the serotypes, <15 are responsible for the majority of infections.

Associated infections
- Respiratory tract: otitis media and sinusitis; *S. pneumoniae* is the most common cause of lower respiratory tract infection.
- It is an important cause of bacterial meningitis, particularly in the elderly.
- Patients who have had splenectomies or whose spleen is functionally impaired have a reduced capacity to produce IgG antibodies to carbohydrate antigens and are particularly susceptible to invasive pneumococcal disease.

Laboratory diagnosis. This is by microscopy and culture of specimens from the infected site, e.g. sputum (Plate 9), CSF (Plate 10), blood. Direct detection of pneumococcal antigen in specimens (CSF, sputum, urine) can be made by an agglutination reaction and is particularly useful in the rapid diagnosis of pneumococcal meningitis and septicaemia. Serological tests for the detection of antibody to pneumococci are of limited use due to the many pneumococcal serotypes.

Treatment and prophylaxis. Until recently, pneumococci have remained sensitive to penicillin and erythromycin. However, penicillin-resistant strains have now emerged and are a particular problem in South Africa and parts of Europe. In some areas of the UK, up to 5% of strains are resistant to penicillin.

A vaccine containing 14 of the most commonly isolated pneumococcal serotypes can be given to patients at particular risk of associated infections, including those with sickle cell disease, asplenia, other forms of immunodeficiency, chronic renal disease, chronic heart disease, chronic lung disease or chronic liver disease. Penicillin prophylaxis may also be given either instead of or in addition to vaccination.

Viridans streptococci

The viridans streptococci are α-haemolytic streptococci; there are a number of different species (e.g. *S. sanguis, S. mutans, S. mitis, S. salivarius, S. bovis*).

Epidemiology. The viridans streptococci are commensals of the human upper respiratory and gastrointestinal tracts. Large numbers are found in the oral cavity.

Morphology and identification. They are facultative aerobic organisms. Colonies show zones of α-haemolysis and are resistant to optochin.

Pathogenicity. Viridans streptococci possess few virulence factors. Attachment to tooth enamel and gums via various carbohydrates is important in establishment and maintenance of colonization. The ability to produce acid particularly by *S. mutans* has been implicated in the development of dental caries.

Associated infections
- *Dental caries*: the factors involved in the pathogenesis of dental caries are complex; the viridans streptococci, particularly *S. mutans*, have been implicated in this process.
- *Bacterial endocarditis*: viridans streptococci are the most common cause of bacterial endocarditis; the organism may enter the blood stream as a result of dental manipulation and attaches to damaged cardiac valves.

Treatment. The viridans streptococci are usually sensitive to penicillin and erythromycin. The treatment of bacterial endocarditis requires antibiotic combinations, e.g. penicillin plus gentamicin.

S. milleri

S. milleri, an important cause of deep-seated abscesses, does not fall neatly into the normal classification of streptococci. It is often α-haemolytic but may be non- or β-haemolytic; strains often group as Lancefield group F, but others may be non-groupable or belong to other Lancefield groups. *S. milleri* is found in the normal human gut and is a common cause of abscesses in the abdominal cavity, chest and brain. It is often found in association with other bacteria.

Culture properties are similar to other streptococci. The organism is sensitive to penicillin but treatment of deep-seated abscesses often requires drainage and additional antibiotics because of the presence of other bacteria.

ENTEROCOCCI

- Until recently, enterococci were classified in the genus *Streptococcus*; the species *S. faecalis* and *S. faecium* were collectively called the faecal streptococci or Group D streptococci. Enterococci have now been designated a separate genus. The enterococci have similar properties to streptococci, but differ in their ability to grow on bile-salt-containing media, e.g. MacConkey's medium.
- They grow well on blood agar, displaying α- or β-haemolysis; some strains are non-haemolytic. The majority of strains belong to Lancefield Group D. The important species are *E. faecalis* and *E. faecium*, which can be distinguished by biochemical tests.
- The principal habitat of enterococci is the gastrointestinal tract. The enterococci cause a number of important infections including endocarditis, urinary tract and wound infections, and abscesses (particularly in association with coliforms and anaerobic bacteria).
- Unlike most streptococci, enterococci, particularly *E. faecium*, have developed resistance to penicillin. The recent emergence of enterococci resistant to glycopeptide antibiotics (e.g. vancomycin) is of particular concern.

ANAEROBIC STREPTOCOCCI

Strictly anaerobic streptococci include the genera *Peptococcus* and *Peptostreptococcus*. They are commensals of the bowel and vagina; and cause abscesses, normally in association with other bacteria. They are sensitive to penicillin and metronidazole.

CHAPTER 7

Clostridia

There are four main clostridial species of medical importance: *Clostridium perfringens*, *C. tetani*, *C. botulinum* and *C. difficile*. Table 7.1 lists infections caused by clostridia.

Definition. Gram-positive bacilli (Plates 2 and 11); strict anaerobes; spore-forming; grow on simple media; many species produce exotoxins.

Epidemiology. Spore formation (Plate 2) allows survival in hostile environments. The ability to survive many forms of heat and chemical disinfection is important in sterilization procedures. Both spores and vegetative forms are found in soil and in the gastrointestinal tracts of mammals.

Classification. Clostridia can be distinguished by morphology, biochemical reactions and antigenic characteristics.

C. perfringens

Morphology and identification
- Gram-positive rod with subterminal oval-shaped spores which form irregular colonies, normally surrounded by a zone of β-haemolysis on blood agar. *C. perfringens* ferments a variety of sugars with the production of gas.
- Five types (A–E) of *C. perfringens* are recognized based on surface antigens and the major types of lethal toxins produced. They are:

  1 type A strains, most commonly found in human infections, producing only the α-toxin;

  2 types B–E, usually associated with disease in various animals, producing α-toxin plus additional toxins.

Pathogenicity. Related to toxin production; the α-toxin is responsible for the toxaemia associated with gas gangrene.

INFECTIONS CAUSED BY CLOSTRIDIA	
Species	**Infection**
C. perfringens	Gas gangrene
	Anaerobic cellulitis
	Food poisoning
C. tetani	Tetanus
C. botulinum	Botulism
C. difficile	Pseudomembranous colitis

Table 7.1 Infections caused by clostridia.

Associated infections
- Often associated with ishaemia (e.g. peripheral vascular disease; diabetes mellitus).
- Anaerobic cellulitis.
- Gas gangrene.
- Food poisoning.

Laboratory diagnosis. Isolation from infected tissue or blood cultures. Identification can be confirmed by the Nagler reaction, which detects α-toxin, a phospholipase. Toxin-producing strains produce a zone of opalescence around growth on egg-yolk-containing media; this effect is inhibited by specific antitoxin to α-toxin (Plate 12).

Treatment and prevention. In the developed world, gas gangrene has largely disappeared as a result of changes in surgical practice with debridement of devitalized tissue. Penicillin prophylaxis is normally prescribed for amputation of the lower limb, particularly when the blood supply is poor, to prevent clostridial infections. Treatment of established gas gangrene involves surgical debridement, high-dose benzylpenicillin, and supportive measures.

C. tetani

Morphology and identification. Gram-positive rod with a spherical terminal spore ('drum stick', Plate 2). They are motile and form a fine film of growth without discrete colonies on blood agar.

Pathogenicity. C. tetani produces a powerful neurotoxin which binds to ganglioside receptors of neurones and blocks neurotransmitter release.

Associated infections. Tetanus may arise when *C. tetani* spores contaminate a wound. Spores germinate under anaerobic conditions (e.g. devitalized tissue), the organism multiplies and produces a powerful neurotoxin, resulting in spastic paralysis.

Laboratory diagnosis. Diagnosis is by direct isolation of the organism from affected tissue, or by detection of circulating toxin by immunoassay.

Treatment and prevention. Although tetanus is rare in the developed world, it remains a major problem worldwide. A toxoid vaccine is available and is important in prevention.

C. botulinum

Morphology and identification. Gram-positive rod, motile with subterminal oval spores. Six types (A–F) are recognized, each producing an antigenically distinct exotoxin.

Pathogenicity. The organism produces a neurotoxin which results in a flaccid paralysis.

Associated infections
- Human botulism is normally caused by types A, B and D. Spores contaminate food (particularly preserved foods) and germinates. The organism then replicates and produces toxin. Ingestion of the toxin results in a flaccid paralysis.
- Infant botulism has been described recently as a cause of flaccid paralysis in infants and is associated with overgrowth of *C. botulism* in the gut.

Laboratory diagnosis. By toxin detection in food or faeces.

Treatment and prevention. Respiratory support and antitoxin. Prevention is dependent on good practice in food preparation and preservation.

C. difficile

C. difficile causes pseudomembranous colitis, a complication of antibiotic therapy.

Morphology and identification. Large Gram-positive bacillus with terminal spores (Plate 13).

Epidemiology. Present in colon of about 20% of healthy individuals; antibiotics may suppress normal flora, allowing overgrowth of *C. difficile* and subsequent toxin production. Person-to-person spread may occur in hospitals, resulting in outbreaks.

Pathogenicity. Produces two toxins; toxin A acts on the gut mucosa whilst toxin B results in a cytopathic effect on tissue culture cells.

Laboratory diagnosis
- The organism can be isolated from faeces by anaerobic culture on selective media. Toxin production is confirmed by exposure of tissue culture cells to culture filtrates of the organism.
- Diagnosis can also be made by direct detection of toxin in faeces:
  (a) incubation of tissue culture cells with faecal filtrates; typical cytopathic effect can be recognized within 48 h; specificity is confirmed by absence of cytopathic changes in the presence of specific antitoxin;
  (b) commercial assays (ELISA, latex agglutination), based on specific antitoxin are available.

Treatment. Oral vancomycin or metronidazole.

CHAPTER 8

Other Gram-positive Bacteria

CORYNEBACTERIA

The genus *Corynebacterium* consists of a number of species which are commensals of man, and one important pathogen, *Corynebacterium diphtheriae*. *C. ulcerans* is also implicated in some infections.

Definition. Gram-positive rods; non-capsulate; non-sporing; aerobic; growth at optimum of 37°C on basic media; growth improved by addition of serum or blood.

C. diphtheriae

C. diphtheriae is the cause of diphtheria, an upper respiratory tract infection with complicating cardiac and neurological pathology.

Morphology and identification
- *C. diphtheriae* is a Gram-positive rod. Incomplete separation of cell walls during division leads to typical 'Chinese letter' morphology.
- When isolating *C. diphtheriae* from throat swabs, a medium containing tellurite is often used; tellurite suppresses growth of many upper respiratory tract commensals, but has no effect on growth of *C. diphtheriae* which produces typical grey-black colonies (Plate 14).
- Three types of colony can be distinguished on tellurite medium. These variants are termed gravis, intermidius and mitis and are associated with severe, moderate and mild clinical symptoms respectively. These colonial variants can also be distinguished by biochemical tests, haemolytic activity and antigenic characteristics.

Epidemiology. Sources of infection are from infected individuals or carriers of toxigenic strains of *C. diphtheriae*. Both nose and throat are sites of carriage and transmission is by droplet spread.

Pathogenicity
• Toxigenic strains of *C. diphtheriae* are not invasive, but multiply locally in the pharynx and produce an exotoxin that destroys epithelial cells, resulting in an acute inflammatory reaction. Toxin may be absorbed into the blood stream, causing damage to myocardial and neural cells.
• A gene coding for toxin production is carried by a temperate bacteriophage and is integrated into the bacterial chromosome.
• Like many bacterial exotoxins, diphtheria toxin has two components: B fragment, responsible for binding to the target cell (B for binding), and A fragment (A for active), responsible for target cell death by inhibition of protein synthesis.

Laboratory diagnosis
• Diagnosis is by direct isolation of the organism from throat swabs, preferably on tellurite medium (Plate 14).
• Toxigenic strains can be confirmed by the Elek test, which is a precipitation reaction in agar with specific antibody to diphtheria toxin.
• Because of the rarity of diphtheria in the developed world, many laboratories no longer screen routinely for the presence of *C. diphtheriae* in throat swabs. It is important to inform the laboratory if a clinical diagnosis of diphtheria is suspected.

Treatment and prevention. Treatment is with diphtheria antitoxin, and penicillin or erythromycin. Contacts should be given antibiotic prophylaxis (erythromycin) and a full vaccination course or booster dose of vaccine, depending on vaccination history. Diphtheria vaccination is part of the routine immunization of children in most developed countries. Travellers to areas of the world where diphtheria is still common should receive a vaccine booster dose.

Other corynebacteria

C. ulcerans, which can be distinguished from *C. diphtheriae* biochemically, is a rare cause of pharyngitis and skin ulcers. Strains producing diphtheria toxin have been isolated, which cause a clinical syndrome identical to that caused by toxin-producing strains of *C. diphtheriae*.

A number of other members of the genus *Corynebacterium* occasionally cause infections. These are part of the skin flora of humans, and are generally referred to as 'diphtheroid bacilli' or 'diphtheroids'. They can cause infections of indwelling intravascular catheters and other prostheses; one particular species, *C. jeikeium*, recognized by its resistance to many antibiotics, appears to be associated particularly with such infections.

BACILLUS

The genus *Bacillus* contains two important human pathogens: *B. anthracis*, the causative organism of anthrax, and *B. cereus*, which is associated with food poisoning.

Definition. Gram-positive bacilli; spore forming; some species capsulate; aerobic; grow over wide temperature range on simple media.

B. anthracis

Epidemiology. *B. anthracis* causes infections principally in animals and is endemic in some areas of the world. Human infections are related to contact with infected animals or animal products (e.g. hides). Spores can survive for long periods in soil.

Morphology and identification. *B. anthracis* has large, Gram-positive, capsulate, non-motile bacilli, which produce oval spores. Identification is confirmed by a variety of morphological and biochemical tests.

Pathogenicity. *B. anthracis* possesses an anti-phagocytic capsule and produces a powerful toxin. The detailed mechanism of action is not yet fully elucidated.

Associated infections. Infection by *B. anthracis* is rare in the UK. It is associated with handling imported animal products.
• Skin infection: an ulcerating skin lesion with necrotic centre (malignant pustule).
• Pulmonary anthrax: spores are inhaled resulting in pulmonary oedema and haemorrhage followed by septicaemia and frequently death.
• Septicaemia: a complication of cutaneous anthrax or pulmonary anthrax.

Laboratory diagnosis. This is by direct isolation from skin lesions or sputum.

Treatment and prevention. Treatment is by administration of penicillin. Prevention is achieved by vaccination of animals and humans at high risk and by slaughter and cremation of infected animals.

B. cereus

B. cereus is an important cause of food poisoning and produces spores, which can survive cooking. The organism is particularly associated with rice dishes.

Pathogenicity is related to the production of two enterotoxins, A and B. Food poisoning by B. cereus can present as either a rapid-onset illness with vomiting, or a diarrhoeal illness.

LISTERIA MONOCYTOGENES

Listeria monocytogenes is an important cause of meningitis and septicaemia in neonates and immunocompromised patients.

Definition. Gram-positive bacilli; motile; non-sporing; non-capsulate; aerobic and facultatively anaerobic; optimum growth at 37°C but will grow slowly at 4–8°C.

Epidemiology. L. monocytogenes is widely distributed in the environment and forms part of the gastrointestinal flora of some animals and, occasionally, humans. It can contaminate a variety of food products (particularly soft cheeses), which may result in outbreaks of food poisoning.

Morphology and identification. L. monocytogenes is β-haemolytic on blood agar and can be distinguished from other Gram-positive bacteria by biochemical tests, and by its characteristic 'tumbling' motility at room temperature in broth cultures.

Pathogenicity. L. monocytogenes is an intracellular pathogen, which can survive within phagocytic cells. Invasive infections may be associated with granuloma formation.

Associated infections
- Congenital/neonatal infection: infection during pregnancy may lead to abortion or premature delivery. L. monocytogenes is an important cause of neonatal meningitis and septicaemia.
- Septicaemia and meningitis may also occur in the immunocompromised, and very occasionally, in immunocompetent patients.

Laboratory diagnosis. By direct isolation of the organism from blood cultures or cerebrospinal fluid.

ERYSIPELOTHRIX

This genus contains only one species: *Erysipelothrix rhusiopathiae*. It is a small, Gram-positive rod, which can be distinguished from *L. monocytogenes* and other Gram-positive bacilli on biochemical characteristics. *E. rhusiopathiae* is associated with a wide range of animals and causes a number of important veterinary infections. In man, the organism causes cellulitis (sometimes termed 'erysipeloid'), particularly in fish handlers, butchers and others involved in handling raw meat and fish.

ACTINOMYCETACEAE

The family Actinomycetaceae has two main genera which are associated with human infections, *Actinomyces* and *Nocardia*.

Definition. Characteristic Gram-positive branching filamentous appearance; non-sporing; grow on simple media; *Actinomyces* are strict anaerobes, *Nocardia* are strict aerobes.

Actinomyces

A. Israeli is the commonest pathogenic species and causes actinomycosis.

Epidemiology and associated infections. The organism is found as a normal commensal in the oral cavity. Infections occur in both the immunocompromised and the normal host. These include:
- abscesses in the oral–facial region often associated with recent trauma or dental extraction;
- abdominal lesions following appendicectomy (rare);
- uterine infections associated with the use of intrauterine contraceptive devices;
- invasive infections in the immunocompromised patient.

Morphology and identification. *A. Israeli* is a Gram-positive bacteria which forms branching filaments. Filaments may aggregate to form visible granules (sulpher granules) in pus. Anaerobic or microaerophilic atmosphere is required for growth which takes up to 10 days before colonies appear. They grow on simple media but growth is enhanced on blood agar.

Treatment. Penicillin for at least 3 months.

Nocardia

Nocardia have features in common with *Actinomyces*; forming Gram-positive filaments, but growing only aerobically. *Nocardia* are found widely in the environment. A number of species are responsible for human infection. *N. asteroides* is an opportunist pathogen causing abscesses in the chest, brain, liver and other organs, particularly in immunocompromised patients. *N. braziliensis* causes similar infections, but mainly in South America. Transmission is usually via the airborne route.

Treatment is often difficult and requires the use of long-term antibiotics, such as sulphonamides.

CHAPTER 9

Gram-negative Cocci

NEISSERIA

The genus *Neisseria* are Gram-negative cocci and include two important pathogens: *Neisseria gonorrhoeae* (gonococcus) and *N. meningitidis* (meningococcus; Table 9.1). Some *Neisseria* species are normal commensals of the human upper respiratory tract.

N. gonorrhoea

Definition. Gram-negative kidney-shaped cocci, usually in pairs (Plate 15); aerobic; optimal growth at 37°C on complex media plus 5% CO_2.

Epidemiology. Obligate human parasite. Asymptomatic females may act as a reservoir; transmission is by sexual contact.

Morphology and identification. Identification is by microscopy when kidney-shaped Gram-negative diplococci are often seen in pus cells. They are oxidase-positive and have strict nutritional requirements for growth. In contrast, many commensal *neisseriae* will grow on simple media. Biochemical reactions, including carbohydrate fermentation, assist identification.

Pathogenicity. Gonococci have cell-surface fimbriae (pili), which are important in adherence to mucosa. They can survive intracellularly. Some strains are able to resist serum lysis. Antigenic variation means multiple infections are common.

Associated infections
- Genital: urethritis, cervicitus; local complications include epididymitis and pelvic inflammatory disease.
- Septicaemia and arthritis (rare).

Treatment and prevention. Penicillins; strains producing penicillinase are

GRAM-NEGATIVE COCCI	
N. meningitidis	Meningitis
	Septicaemia
N. gonorrhoea	Urethritis, cervicitis
	Epididymitis
	Pelvic inflammatory disease
	Neonatal ophthalmia
	Septicaemia
	Arthritis
Moraxella catarrhalis	Lower respiratory tract infections

Table 9.1 Gram-negative cocci and associated infections.

susceptible to cephalosporins (e.g. cefuroxime) or spectinomycin. No vaccine available.

Laboratory diagnosis. Specimens should be rapidly transported to the laboratory as gonococci die readily on drying. Diagnosis is by microscopy and culture of pus and secretions from various sites (depending upon the infection): cervix, urethra, rectum, conjunctiva, throat, or synovial fluid. Gram stain of urethral and cervical discharge smears for Gram-negative diplococci (Plate 15). Clinical samples are plated onto enriched selective media. Identification is confirmed by carbohydrate fermentation tests.

Blood cultures are useful in diagnosing disseminated infection, but serology is rarely used because of poor specificity and sensitivity.

N. meningitidis

Definition. Gram-negative kidney-shaped cocci, usually in pairs; aerobic; optimum growth at 37°C on complex media in air with 5% CO_2.

Epidemiology
- *N. meningitidis* is an obligate human parasite causing meningitis and septicaemia; man is the only natural host. The organism is carried in the upper respiratory tract by some individuals. Sporadic cases or epidemics occur worldwide.
- At least 13 serogroups of meningococci, based on capsular polysaccharides have been identified, the most important being groups A, B, C, W-135 and Y.
- Patients with defects of the later components of the complement system (C8, C9) may suffer recurrent meningococcal infections.

Morphology and identification
- Identification is by microscopy when kidney-shaped, Gram-negative cocci are often seen within polymorphs. *N. meningitidis* is oxidase-positive and biochemical reactions (carbohydrate fermentation reactions) are used for identification.
- Serogrouping is performed by reference laboratories to provide epidemiological information.

Pathogenicity. A polysaccharide capsule protects against phagocytosis and promotes intracellular survival. Cell-wall endotoxin is important in the pathogenesis of severe meningococcal disease.

Associated infections
- Meningitis, often with septicaemia.
- Septicaemia without meningitis (less common).

Laboratory diagnosis
- Specimens of blood and cerebrospinal fluid (CSF) for microscopy, culture and antibiotic sensitivity; nasopharyngeal swabs are useful to determine carriage.
- Direct detection of meningococcal antigen in CSF or urine by agglutination reaction.
- Serology: antibodies to meningococcal polysaccharide antigens are assayed occasionally to provide a retrospective diagnosis.

Treatment. Penicillin; vaccine available for serogroups A and C.

OTHER GRAM-NEGATIVE COCCI

Other *Neisseria*, such as *N. subflava* and *N. lactamica*, are part of the normal flora of the upper respiratory tract, and rarely cause disease. *Moraxella catarrhalis* (formerly *Branhamella catarrhalis*) is also a member of the commensal flora of the upper respiratory tract and is an occasional cause of lower respiratory tract infections, particularly exacerbations of chronic bronchitis. Most *M. catarrhalis* strains produce β-lactamase which breaks open the β-lactam ring of penicillin. β-lactamase-stable β-lactam agents, such as cefuroxime or co-amoxiclav, are therefore required for therapy.

CHAPTER 10

Enterobacteriaceae

Enterobacteriaceae are a heterogeneous group of Gram-negative aerobic bacilli, which are commensals of the intestinal tract of mammals. They are also referred to as coliforms or enteric bacteria. The family includes the following important genera: *Escherichia, Enterobacter, Klebsiella, Proteus, Serratia, Shigella* and *Salmonella* (Table 10.1).

Definition. Gram-negative bacilli (Plate 16); aerobic and facultatively anaerobic growth; optimal growth normally at 37°C (Plate 17); grow readily on simple media; ferment wide range of carbohydrates; oxidase-negative; some are motile; bile tolerant and grow readily on bile-salt-containing media, e.g. MacConkey's agar.

Morphology and identification
- Fermentation of lactose to produce pink colonies on MacConkey's agar is a property of the genera *Escherichia, Enterobacter, Klebsiella* and *Serratia*. The genera *Salmonella, Shigella* and *Proteus* do not ferment lactose and form pale colonies on MacConkey's agar.
- Various biochemical tests are used for identification:
    (a) carbohydrate fermentation reactions;
    (b) production of urease, which splits urea with release of ammonia;
    (c) hydrogen sulphide production; amino acid decarboxylation and indole production.
- Enterobacteriaceae possess a variety of antigens: somatic or cell wall ('O'), flagella ('H') and capsular ('K'). These are used to subdivide some genera and species.

ESCHERICHIA

The genus *Escherichia* contains only one species, *E. coli*.

Epidemiology and associated infections. Although *E. coli* is a commensal of the human intestine, it can cause a variety of important infections: includ-

ENTEROBACTERIACEAE

Genus/species	Lactose fermentation	Colonies on MacConkey's agar	Common infections
Escherichia coli	+	Pink	Urinary tract infection, diarrhoeal disease, wound infections
Klebsiella	+	Pink	Hospital-acquired infection: urinary tract infection
Enterobacter	+	Pink	pneumonia
Serratia	+	Pink	wound infection
Salmonella typhi and Salmonella paratyphi	−	Pale	Enteric fever
Other salmonellae	−	Pale	Enteritis Septicaemia
Shigella	−*	Pale	Enteritis
Proteus	−	Pale	Urinary tract infection
Yersinia	−	Pale	Y. enterocolitica, enteritis, Y. pestis, plague, Y. pseudotuberculosis, mesenteric adenitis

* Shigella sonnei ferments lactose, but only after incubation >24 h.

Table 10.1 Enterobacteriaceae: main characteristics and infections.

ing infections of the gastrointestinal tract, urinary tract, bilary tract, lower respiratory tract and septicaemia.

Pathogenicity
- Fimbriae facilitate adherence to mucosal surfaces of the intestinal and urinary tracts.
- As with all Gram-negative bacteria, E. coli have lipopolysaccharides (endotoxin) in their cell walls. Endotoxin is liberated when Gram-negative bacteria lyse, resulting in complement activation, intravascular coagulopathy and endotoxic shock.
- Exotoxins are responsible for the diarrhoeal diseases caused by enteropathogenic E. coli.
- E. coli has capsular antigens; the K1 antigen is associated with neonatal meningitis.

Antibacterial therapy. E. coli is commonly resistant to penicillin and ampicillin (by production of β-lactamases). Antibiotics often used to treat *E. coli* infections include the cephalosporins, trimethoprim and aminoglycosides; strains isolated from hospitalized patients are often more resistant to antibiotics.

KLEBSIELLA

The genus *Klebsiella* contains a number of species, including *K. pneumoniae*, *K. aerogenes* and *K. oxytoca*; these are separated on the basis of biochemical tests. Pathogenicity is associated with capsule production; there are many capsular serotypes.

Klebsiellae are widespread in the environment and in the intestinal flora of man and other mammals. Infections are often opportunist and associated with hospitalization, particularly in high-dependency units. They include pneumonia, urinary tract and wound infections and neonatal meningitis. Outbreaks of nosocomial (hospital-acquired) infections occur.

Klebsiellae often produce β-lactamases and are resistant to ampicillin. Cephalosporins, β-lactamase-stable penicillins and aminoglycosides are commonly used to treat *Klebsiella* infections, but multiply resistant strains may limit antibiotic choice.

ENTEROBACTER AND SERRATIA

The genera *Enterobacter* and *Serratia* are closely related to klebsiellae. Infections occur principally in hospitalized patients and include the lower respiratory and urinary tracts. Hospital cross-infection with antibiotic-resistant strains is a particular problem.

SALMONELLA

The genus *Salmonella* contains a large number of species (more correctly, serotypes; see below). *S. typhi* and *S. paratyphi* cause enteric fever (typhoid or paratyphoid); other salmonellae cause enteritis.

Classification
- Classification of the genus *Salmonella* is complex, with >2000 serotypes being distinguished. Many of these have been given binomial names, e.g. *S. typhimurium* and *S. enteritidis*, although they are not separate species,

merely different serotypes. In clinical practice, laboratories identify organisms according to their binomial name.
- Salmonellae have both H (flagellar) and O (somatic) antigens. There are over 60 different O antigens, and individual strains may possess several O antigens and H antigens; the latter can exist in different forms, termed 'phases'. *S. typhi* has a capsular-associated antigen referred to as 'Vi' (for virulence) which is related to invasiveness.
- Agglutination tests with antisera for different O and H antigens form the basis for the serological classification of the salmonellae. Further strain differentiation of salmonellae for epidemiological purposes can be obtained by phage typing.

Epidemiology. Salmonellae are commensals of many animals including poultry, domestic pets, birds and humans. Transmission is via the faecal-oral route. The infective dose is relatively high (approximately 10^6 organisms) and multiplication in food is important for effective transmission. A chronic carrier state can occur.

Morphology and identification
- Salmonellae are motile and produce acid, and occasionally gas, from glucose and mannose.
- They are resistant to sodium deoxycholate which inhibits many other Enterobacteriaceae and deoxycholate agar is used as a selective media used to isolate salmonellae from stool specimens.
- Salmonellae do not ferment lactose and form pale colonies on MacConkey medium; this aids recognition of salmonellae colonies in mixed cultures.
- Further biochemical tests are required for definitive identification. Serotyping by slide agglutination is used for speciation.

Pathogenicity. Salmonellae invade the gut resulting in an inflammatory response and subsequent diarrhoea. The 'Vi' antigen of *S. typhi* is associated with invasiveness.

Associated infections
- Enterocolitis; rarely associated with septacaemia, osteomyelitis, septic arthritis or abscesses.
- Enteric fever; caused by *S. typhi* and *S. paratyphi*. Septicaemic illness.

Laboratory diagnosis
- Enterocolitis: culture of stool samples; identification of salmonellae by

biochemical and agglulination tests. Phagetyping can be used for typing individual strains.

- Enteric fever: isolation of *S. typhi* or *S. paratyphi* from blood cultures (first week of infection), from urine (second week), or from faeces (first week onwards). Serology (Widal test) is now rarely performed because of unreliable results.

Treatment
- Enteric fever: various antibiotics including chloramphenicol, trimethoprim and ciprofloxacin.
- Enterocolitis: self-limiting; antibiotics (e.g. ciprofloxacin) reserved for severe septacaemic cases particularly in the elderly, very young and the 'immunocompromised'.

SHIGELLA

Classification. The main pathogenic species are *S. sonnei*, *S. boydii*, *S. dysenteriae* and *S. flexneri*. They are distinguished by biochemical reactions and antigenic characteristics ('O' antigens).

Epidemiology. Obligate human pathogens with no animal reservoirs; transmissions via faecal-oral route with low infective dose (10^3 organisms). Direct person-to-person spread is common; chronic carrier state is rare.

Morphology and identification. Shigellae are non-motile (they have no flagella). They are resistant to sodium deoxycholate and so will grow on deoxycholate agar (see salmonellae). They are non-lactose or late lactose (*S. sonnei*) fermenters. Further biochemical tests are carried out for definitive identification and serotyping by slide agglutination is used for speciation.

Pathogenicity. Shigellae cause disease by invasion and destruction of the colonic mucosa. They also produce an exotoxin.

Associated infections. Dysentery (diarrhoea with blood and pus); septicaemic complications are extremely rare.

Laboratory diagnosis. Stool culture on selective media.

Treatment. Often self-limiting; antibiotics (e.g. trimethoprim, chloramphenicol, ciprofloxacin) reserved for severe cases (often caused by *S. dysenteriae*).

PROTEUS

The genus *Proteus* contains a number of species, e.g. *Proteus mirabilis* and *P. vulgaris*. Characteristics include:
- non-lactose fermenting, produce pale colonies on MacConkey's agar;
- motile, tendency to 'swarm' on blood agar (Plate 18);
- important cause of urinary tract and occasionally abdominal wound infections.

YERSINIA

The genus *Yersinia* contains three human pathogens: *Yersinia pestis*, *Y. pseudotuberculosis* and *Y. enterocolitica*; these species are identified by biochemical tests.

Y. pestis

Y. pestis is a cause of plague (Black Death). Although mainly of historical interest in Europe, plague remains endemic in some areas of the world.

It is primarily a pathogen of rodents and is transmitted to humans via infected fleas; lymph nodes associated with the flea bite enlarge to form a bubo (bubonic plague). Septicaemia and pneumonia may follow (pneumonic plague). Person-to-person spread via droplets occurs in pneumonic plague. Treatment for *Y. pestis* infection is with tetracycline.

Y. pseudotuberculosis

This is primarily an animal pathogen but in humans it is an occasional cause of mesenteric adenitis, rarely septicaemia. It is probably transmitted to humans via contaminated food.

Y. enterocolitica

Y. enterocolitica causes diarrhoeal disease, terminal ileitis and mesenteric adenitis. Infection may be complicated by septicaemia or reactive polyarthritis. Laboratory diagnosis is by culture of stool or blood specimens or by serology.

CHAPTER 11

Parvobacteria

The parvobacteria are a group of genera (Table 11.1). All are Gram-negative coccobacilli, and are often described collectively because of common characteristics. They are relatively small (0.5–1.0 μm) and grow only on enriched media. The group includes the genera *Haemophilus*, *Bordetella*, *Legionella*, *Pasteurella* and *Francisella*.

HAEMOPHILUS

Definition. Gram-negative bacilli; fastidious growth requirements; non-motile; some strains capsulate; aerobic and anaerobic growth.

H. influenzae

Classification. Strains may be capsulate or non-capsulate. Capsulate strains are divided into six serotypes (designated b–f) on the basis of capsular antigens; type b strains are an important cause of invasive infection, particularly in small children.

Epidemiology. Non-capsulate strains of *H. influenzae* are common commensals in the upper respiratory tract. Capsulate strains, including *H. influenzae* type b, are also carried in a small number of healthy individuals ('carriers') and are an important reservoir for invasive disease.

Morphology and identification
- *H. influenzae* is nutritionally demanding and grows only on enriched media requiring haemin (factor X) and nicotinamide-adenine dinucleotide (factor V) for growth. Simple nutrient agar contains no X or V factor and strains of *H. influenzae* will grow only around paper discs containing both these factors (Plate 19).
- Growth is poor on blood agar which contains factor X but only small amounts of available factor V; heat treatment of blood before incorporation in agar produces a medium known as chocolate agar which contains larger quantities of both factors and allows growth of *H. influenzae*. Some

PARVOBACTERIA		
Genus	**Species**	**Associated infections**
Haemophilus	H. influenzae	Epiglottitis (capsulate, type b); invasive infections, including meningitis in children <5 years (capsulate, type b) Exacerbations of chronic lung disease (non-capsulate)
	H. parainfluenzae	Exacerbations of chronic lung disease.
	H. ducreyi	Genital ulcers (chancroid)
Brucella	B. melitensis B. abortus B. suis	Brucellosis
Pasteurella	P. multocida	Cellulitis following animal bites
Legionella	L. pneumophilia	Pneumonia, Pontiac fever
Bordetella	B. pertussis	Pertussis ('whooping cough')

Table 11.1 Common infections associated with parvobacteria.

bacteria, particularly staphylococci, produce factor V and *H. influenzae* strains can sometimes be seen growing around staphylococcal colonies on blood agar.

Pathogenicity. The polysaccharide capsule of certain strains, particularly type b, is a major virulence factor. Antibodies to the type b capsule are protective for invasive *H. influenzae* type b infections.

Associated infections
• *H. influenzae* type b strains are an important cause of serious infection in young children (aged 6 months to 5 years). Outside this age group, infections are less common.
• Infections include meningitis, epiglottitis, cellulitis, otitis media, osteomyelitis and septic arthritis, and pneumonia.
• Non-type b capsulate strains occasionally cause invasive infections. Non-capsulate strains are important respiratory pathogens, particularly as a cause of acute exacerbations of chronic lung disease. Their role in primary lobar pneumonia is unclear.

Laboratory diagnosis
• Diagnosis is by direct isolation of the organism from infected sites, e.g. cerebrospinal fluid (CSF) and blood. Identification is made by assessing X

and V dependency and biochemical tests. Serotyping is carried out on capsulate strains when isolated from invasive infections.
- Direct detection of capsular antigen in CSF can be made for the rapid diagnosis of *H. influenzae* type b meningitis.

Treatment and prophylaxis
- Many *H. influenzae* strains produce β-lactamase and are resistant to ampicillin. Until recently, chloramphenicol was the agent of choice for treating serious infections with *H. influenzae* type b. However, in some countries, chloramphenicol resistance is increasing and cefotaxime is used as first-line treatment. Although chloramphenicol resistance is relatively uncommon in the UK, many clinicians now select cefotaxime.
- A vaccine to the type b polysaccharide capsule has been developed and introduced recently into the routine immunization schedule for infants in the UK; this has led to a rapid reduction in the number of invasive *H. influenzae* type b infections. Rifampicin is used for the prophylaxis of contacts of cases of *H. influenzae* type b meningitis, and as an adjunct to therapy to reduce nasopharyngeal carriage.

H. parainfluenzae

H. parainfluenzae is non-capsulate and differs from *H. influenzae* in requiring only factor V for growth. It is a common upper respiratory tract commensal, but its role in exacerbations of chronic bronchitis and upper respiratory tract infection is ill-defined.

H. ducreyi

H. ducreyi causes genital ulcers (chancroid), a sexually transmitted disease seen particularly in the tropics.

BRUCELLA

The genus *Brucella* contains three species responsible for human infections: *Brucella melitensis*, *B. abortus* and *B. suis*.

Definition. Small, Gram-negative cocco-bacilli; non-motile; non-capsulate; grow slowly aerobically with the addition of 10% CO_2; require enriched media.

Epidemiology. Human brucellosis is a zoonotic infection, the main reservoirs being goats (*B. melitensis*), cattle (*B. abortus*) and pigs (*B. suis*). In the

UK, *B. abortus* is the most common species isolated whilst *B. melitensis* is found most commonly in the Mediterranean area. Infection is associated with close contact with farm animals, e.g. farm workers, veterinary surgeons, or through ingestion of unpasteurized cows' or goats' milk.

Morphology and identification. *Brucella* species are aerobic and require the presence of CO_2 for growth. The organism only grows slowly. Identification is based on morphology, growth characteristics and biochemical reactions. The three species of *Brucella* may be distinguished by biochemical reactions.

Pathogenicity. *Brucella* species are intracellular pathogens and can survive the intracellular killing mechanisms of phagocytic cells. Infection with *Brucella* is typical of an intracellular pathogen with the formation of multiple granulomatous lesions in several organs.

Associated infection. Infection with *Brucella* is associated with fever and a variety of non-specific symptoms, such as malaise and weight loss.

Laboratory diagnosis
• Diagnosis is made by direct isolation from blood cultures or culture of aspirated bone marrow after prolonged incubation (2–3 weeks). *Note:* routine blood cultures are incubated normally for only 3–7 days and it is important to inform the laboratory if brucellosis is suspected.
• Serological tests are available; these aid diagnosis as many patients present late in their illness.

Treatment. Prolonged treatment is necessary (>2 weeks) with tetracycline or co-trimoxazole. This reflects the intracellular location of the organisms.

LEGIONELLA

Legionella pneumophilia and other closely related *Legionella* species are a cause of lower respiratory tract infections in humans. The organism was first recognized following investigations into an outbreak of respiratory illness in delegates at an American Legion Convention in Philadelphia, USA, in 1976.

Definition. Gram-negative bacilli; fastidious growth requirements; non-capsulate; motile.

Epidemiology. The organism is found in water systems, particularly heating systems and water-cooled air-conditioning plants. Although relatively fastidious in its growth requirements, the organism grows in water at temperatures between 20 and 45°C; this may be related to growth within amoebae which are found naturally in the same environment or to its ability to survive in biofilms.

Pathogenicity. Legionellae can survive and grow within phagocytic cells neutralizing the normal bacteria-killing mechanisms.

Associated infections. *L. pneumophila* can cause a range of respiratory illnesses, from atypical pneumonia to a flu-like illness without lower respiratory symptoms, called Pontiac fever.

Laboratory diagnosis
- Diagnosis is made by culture of sputum or broncheolar-alveolar lavage fluid on specialized media; colonies take about 3 days to grow. Legionellae are classified into species according to surface antigens.
- Legionellae can be detected directly in respiratory specimens by monoclonal antibodies labelled with fluorescein.
- Serological tests are often used to confirm the diagnosis of legionellosis; a rise in serum antibody titre may take up to 6 weeks.

Treatment. This is by administration of erythromycin combined, in severe cases, with rifampicin.

BORDETELLA

The genus *Bordetella* contains four species; *B. pertussis*, the cause of whooping cough, is the only important human pathogen.

Definition. Small coccobacilli; obligate aerobes; fastidious growth requirements; non-motile; non-capsulate.

Epidemiology. *B. pertussis* is a human pathogen with no animal reservoir. Transmission is by droplet spread from infected persons.

Pathogenicity. A variety of virulence factors have been described, including adhesins, toxins and haemolysins.

Associated infections. Whooping cough (pertussis).

Laboratory diagnosis. Culture of perinasal swabs on charcoal blood agar. Colonies may take 2–3 days to grow.

Treatment and prophylaxis. Erythromycin is given to reduce infectivity. If given during the incubation period, it may reduce clinical symptoms. Immunization during childhood is widely used.

PASTEURELLA

Pasteurella multocida is the only common human pathogen in this genus. The organism is primarily a pathogen of animals and is found as part of the normal oral flora of cats and dogs. Human infections arise following animal bites and result in a local infection with surrounding cellulitis; treatment is with penicillin. Laboratory diagnosis is by culture of the organism, which grows well on blood agar.

CHAPTER 12

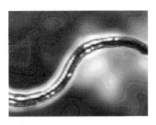

Campylobacter, Helicobacter and Vibrios

CAMPYLOBACTER

Definition. Curved, Gram-negative bacilli; motile with a polar flagellum; non-sporing; capsulate; some species show optimal growth at 42°C; require microaerophilic atmosphere plus 10% CO_2.

Classification. There are several species, including *Campylobacter jejuni, C. coli* and *C. fetus*.

Epidemiology
- Large resevoir of campylobacters in cattle, poultry, wild birds and other animals including pets.
- Transmission via faecal-oral route often by consumption of contaminated food (e.g. chicken, milk).

Morphology and identification
- Campylobacters have a characteristic morphology.
- They grow on selective media in a microaerophilic atmosphere plus 10% CO_2 at 42°C. Several selective media are available for culture of the organism, including Skirrow's medium, which contains the antibiotics vancomycin, polymyxin B and trimethoprim. Colonies are colourless or grey with a metallic sheen.
- Biochemical tests can be used for identification as campylobacters are oxidase- and catalase-positive. Further biochemical tests are used for speciation.

Pathogenicity. Campylobacter cell walls contain endotoxin. Cytopathic extracellular toxins and enterotoxins have also been demonstrated, but their exact role in pathogenicity is unclear.

Associated infections. Acute enteritis, rarely complicated by septicaemia.

Laboratory diagnosis
- Culture of stool sample on selective media at 42°C for 42–72 h.
- Serology may be undertaken but detection of serum antibodies to campylobacters is rarely used in practice.

Treatment. Antibiotics often not required; erythromycin for severe cases, particularly in the elderly.

HELICOBACTER

H. pylori

Definition. Curved Gram-negative bacilli; motile; non-sporing; non-capsulate.

Classification. H. pylori, first cultured in 1982, was originally classified as *Campylobacter pyloridis*, but has subsequently been assigned to the genus *Helicobacter*.

Epidemiology
- H. pylori can survive closely attached to human gastric epithelial cells below the protective mucus layer.
- The organism is found most frequently in older patients. In the developed countries, it is present infrequently in children <10 years (<1%), but is found in approximately 50% of the population at 50 years of age.
- Transmission is thought to occur by the faecal–oral or oral–oral routes and is associated with close contact (intrafamilial) and poor sanitation.

Morphology and identification
- H. pylori grow at 37°C on enriched media (e.g. chocolate agar plates) in a microaerophilic atmosphere; colonies appear after 3–4 days.
- They possess four to six unipolar sheathed flagella (Plate 20); campylobacters have only a single polar unsheathed flagella and produce catalase, oxidase, and urease; the test for urease activity is particularly important in laboratory identification.

Pathogenicity and associated infections. There is a strong association between the presence of *H. pylori* gastritis and gastric and duodenal ulceration. There is also increasing evidence of a link between *H. pylori* and gastric carcinoma. Various pathogenic mechanisms have been proposed for *H. pylori*, but it is not yet clear which are important *in vivo*.

Laboratory diagnosis
- Culture of gastric biopsy specimens and serological tests for specific serum immunogammaglobulin (IgG) antibodies.
- A biopsy urease test may be undertaken. Gastric biopsy specimens may be examined directly for the presence of urease produced by *H. pylori* (growth of the organism is not required). Any urease present hydrolyses urea in a broth to produce ammonia; the rise in pH is detected by an indicator. This test is specific as *H. pylori* is unique in producing a urease with a high affinity and rate of activity. The test is simple, rapid and relatively inexpensive; however, some positives are missed.
- A breath test is also available. ^{13}C- or ^{14}C-labelled urea is given to the patient; if *H. pylori* is present, the urease splits the urea, producing labelled CO_2 which is detected in expired air by various methods.

Treatment. Clearance of *H. pylori* with amoxycillin, metronidazole and bismuth salts, or H_2 proton pump inhibitors, such as omeprazole, has resulted in the resolution of the symptoms and signs in over 80% of individuals. Combinations of other antimicrobial agents, including clarithromycin, have also been used, with similar success. The use of ranitidine bismuth citrate with metronidazole and clarithromycin appears to prevent emergence of antibiotic resistance which is becoming an increasing problem.

VIBRIOS

Vibrios are comma-shaped bacteria; the main species of medical importance are *Vibrio cholerae*, *V. parahaemolyticus*, *V. vulnificus* and *V. alginolyticus*.

V. cholerae

Definition. Gram-negative; comma-shaped bacteria; aerobic and facultatively anaerobic; grow in alkaline conditions; motile; ferment carbohydrates; oxidase-positive.

Classification
- *V. cholerae* strains are subdivided according to O-antigens. *V. cholerae* 01 is the cause of cholera. Other *V. cholerae* (non-01) strains may occasionally cause diarrhoea.

- *V. cholerae* 01 has a number of biotypes including the 'classic' strain, which was the principal cause of cholera until the mid-1960s. The 'eltor' biotype has been responsible for most cases of cholera during the last two decades.

Epidemiology
- Found in water contaminated with human faeces; no animal reservoirs.
- Important cause of severe infection in developing countries where potable water and sewage systems are poor.

Morphology and identification
- *V. cholerae* are characterized by an ability to grow in alkaline conditions (pH >8.0). Alkaline broth is used to grow the organism selectively from faecal samples. A special selective medium (thiosulphate–citrate–bile salt–sucrose agar) is also used. *V. cholerae* form characteristic yellow colonies on this medium.
- Specific slide agglutination reactions for 'O' antigens distinguish *V. cholerae* from non-cholera vibrios.
- Biochemical tests are used for confirmation.

Pathogenicity. *V. cholerae* produce an enterotoxin which acts on intestinal epithelial cells, stimulating adenyl cyclase activity. This results in water and sodium ions passing into the gut lumen to produce profuse, watery diarrhoea.

Associated infections. Acute enteritis.

Laboratory diagnosis. Isolation of organism from faeces on selective media.

Treatment. Rehydration; tetracycline.

Non-cholera vibrios

V. cholerae which do not agglutinate with 01 antisera are known as non-cholera vibrios. These organisms may also cause diarrhoea, which is normally much less severe than classic cholera.

V. parahaemolyticus

V. parahaemolyticus is a cause of food poisoning 12–18 h following the ingestion of contaminated seafood, particularly oysters. The illness is characterized by vomiting and diarrhoea.

CHAPTER 13

Pseudomonas and other Aerobic Gram-negative Bacilli

PSEUDOMONAS

Members of the genus *Pseudomonas* ('pseudomonads') are widely distributed in nature; *P. aeruginosa* is the most important pathogenic species in man, although other species, particularly *P. cepacia*, can cause infections.

Definition. Gram-negative bacilli; strict aerobes; grow on simple media at wide temperature range; non-motile; non-capsulate; some species pigmented.

P. aeruginosa

Epidemiology
- *P. aeruginosa* is an important organism in hospital-acquired infection. It is a normal commensal in the human gastrointestinal tract, but may colonize other sites when host defences are compromised, including burns and leg ulcers, the respiratory tract of patients with cystic fibrosis or bronchiectasis, and the urinary tract of patients with long-term indwelling urethral catheters.
- *P. aeruginosa*, like other members of the genus, has the ability to grow with minimal nutrients, for example, in water and in the presence of some disinfectants; these properties are the key to its important role as a hospital pathogen.

Morphology and identification. The organism is a strict aerobe which grows on most media, producing a characteristic greenish pigment. It can be distinguished from the Enterobacteriaceae by its oxidative metabolism (oxidase-positive) and the inability to grow anaerobically. The different species of the genus *Pseudomonas* can be distinguished by biochemical tests.

Pathogenicity. *P. aeruginosa* is relatively non-pathogenic; it characteristically causes infections in hospitalized patients, particularly those who are immunocompromised. The organism produces several enzymes which allow spread through tissues (elastase) and a protease which breaks down IgA on mucosal surfaces.

Associated infections. *P. aeruginosa* principally causes opportunist infections including:
- skin infections: associated with burns and venous ulcers;
- chest infections: in patients on intensive care units and patients with cystic fibrosis;
- urinary tract infections: particulary with associated long-term urethral catheterization;
- septicaemia;
- eye infections;
- otitis externa.

Laboratory diagnosis. Isolation of the organism from relevant body sites. The isolation of *P. aeruginosa* from several patients may suggest hospital cross-infection. In such circumstances, strains of *P. aeruginosa* can be further characterized by pyocin typing, serotyping or molecular biological techniques.

Treatment. One of the characteristics of *P. aeruginosa* is resistance to antibiotics; some newer antibiotics have been designed specifically to combat *P. aeruginosa*. Clinically important anti-pseudomonal antibiotics include:
- aminoglycosides (e.g. gentamicin);
- broad-spectrum penicillins (e.g. piperacillin);
- third-generation cephalosporins (e.g. ceftazidime);
- quinolones (e.g. ciprofloxacin).

On hospital units where antibiotics are used frequently (e.g. special care baby units, intensive care units), *P. aeruginosa* isolates may become resistant to these antibiotics. Isolation of patients colonized by multiresistant *P. aeruginosa* strains is an important part of controlling hospital infections.

P. cepacia

This member of the genus has recently been recognized as an important cause of lower respiratory tract infection in patients with cystic fibrosis.

P. maltophilia

Recently reclassified as *Stenotrophomonas maltophilia*, this organism is a cause of hospital-acquired infections and is often multiply antibiotic resistant.

ACINETOBACTER

Acinetobacter species share a number of common features with pseudomonads. They are short, Gram-negative bacilli; strict aerobes and grow well on simple media with minimal nutrients. However, unlike pseudomonads they are oxidase negative. They occur in many environments, including hospitals, and are an important cause of hospital-acquired infection. Hospital outbreaks, particularly chest infections, occur particularly in intensive care units. *Acinetobacter* species frequently acquire plasmids expressing multiple antibiotic resistance patterns.

OTHER GRAM-NEGATIVE BACTERIA

Capnocytophaga

This is a Gram-negative bacillus which grows slowly (>5 days) on complex media and has an absolute requirement for carbon dioxide. It is a normal commensal of the oral cavity. It causes septicaemia in patients with immunodeficiency, particularly when oral mucosal ulceration is also present, and is a rare cause of infection in immunocompetent patients (e.g. lung abscess, endocarditis).

Cardiobacterium hominis

C. hominis is a Gram-negative bacillus with tapering ends. It has a tendency to form long filaments. It grows best on enriched media in a microaerophilic environment with additional CO_2. *C. hominis* is a normal commensal of the human upper respiratory tract and is a rare cause of bacterial endocarditis.

Chromobacterium violaceum

This is a Gram-negative bacillus which grows readily on simple media to produce typical violet-pigmented colonies. The organism is saprophytic and is found in soil and water, particularly in tropical and subtropical areas. It may cause pneumonia, septicaemia and abscesses.

Eikenella corrodens

E. corrodens is a Gram-negative bacillus which forms typical pitting colonies. It grows aerobically and anaerobically on complex media; colonies may take 5 days to appear. The organism is found as a normal commensal of the human upper respiratory tract and may be isolated from a variety of infections, including dental abscesses, soft-tissue infections of the neck, pulmonary and abdominal abscesses, wound infections following human bites, normally in association with other organisms such as anaerobes, coliforms and streptococci. *E. corrodens* is an occasional cause of endocarditis.

Gardnerella vaginalis

G. vaginalis is a Gram-negative rod which occasionally appears Gram-positive ('Gram-variable'). It grows aerobically and anaerobically on complex media. It is a normal commensal of the vaginal flora, but is associated with bacterial vaginosis, a condition characterized by an 'ammonia smelling' vaginal discharge and the presence of clue cells (Plate 21) (Epithelial cells with attached *G. vaginalis*.)

Streptobacillus moniliformis

S. moniliformis is a short Gram-negative rod which may also appear as L-forms, and grows slowly on complex media. A commensal of healthy rats, it is one cause of rat bite fever in man (the other aetiological organism being *Spirillum*), an infection characterized by fever, vomiting, headache, arthralgia and rash. *S. moniliformis* is also the cause of 'streptobacilliary fever', sometimes called Haverhill fever, after the first recorded epidemic of the infection at Haverhill, USA. Outbreaks have occurred in the UK and may be associated with the consumption of unpasteurized milk.

CHAPTER 14

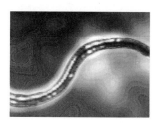

Gram-negative Anaerobic Bacteria

The Gram-negative anaerobes include the genera *Bacteroides*, *Fusobacterium* and *Leptotricia*. Unlike the Gram-positive anaerobes clostridia they do not form spores. Gram-negative anaerobes form a major component of the normal flora of man.

Definition. Gram-negative bacilli; non-motile; non-sporing; strict anaerobes; grow on complex media; some species capsulate.

BACTEROIDES

The genus *Bacteroides* contains a large number of species, the most important being *B. fragilis*.

Epidemiology. *Bacteroides* make up the largest component of the bacterial flora of the human intestine, and are an important part of the normal flora of the mouth and vagina. *Bacteroides* protect against colonization and infection by more pathogenic bacteria; they can also cause infections.

Morphology and identification. Gram-negative bacilli; occasionally form short filaments. Some species have a polysaccharide capsule. They are strictly anaerobic organisms which grow on complex media; visible colonies appear only after 48 h incubation. Some species produce a pigment which results in black colonies. (*Note:* recent taxonomic changes have placed black-pigmented *Bacteroides* in a separate genus, *Prevotella*; Plate 22).

Associated infections. These include: intra-abdominal sepsis (wound infections, abscesses, peritonitis) often in mixed infections with coliforms; abscesses in other organs, particularly brain (again, often mixed with coliforms); oral infections, e.g. acute ulcerative gingivitis; and aspiration pneumonia and lung abscesses. *Bacteroides* may have a role in non-specific vaginosis.

Treatment
- Post-operative, intra-abdominal infections can be prevented by good surgical technique and prophylactic antibiotics.
- Metronidazole and clindamycin are effective antibiotics against *Bacteroides*; most species are sensitive to penicillin but some, particularly *B. fragilis*, are resistant because of β-lactamase production.

FUSOBACTERIUM

Fusobacterium are spindle-shaped, Gram-negative bacilli with pointed ends. They are strict anaerobes and are often nutritionally demanding. The genus is classified into various species, *F. necophorum* and *F. nucleatum* being the most important in human infections.

Fusobacteria are commensals of the oral cavity. They are one of the causative organisms, along with other anaerobes, of acute ulcerative gingivitis (primarily *F. nucleatum*) and other oral infections. *F. necophorum* may occasionally cause septicaemia with metastatic abscesses.

LEPTOTRICIA

Leptotricia are long, slightly curved rods with pointed ends. The genus consists of a single species, *L. buccalis*, which is part of the normal oral flora and may be associated with oral infections such as Vincent's angina.

CHAPTER 15

Treponema, Borrelia and Leptospira

SPIROCHAETES

The bacteria in the order Spirochaetales, commonly referred to as spirochaetes, are grouped together because of common morphological properties. They are thin (0.1–0.5 × 5.0–20.0 μm), helical, Gram-negative bacteria. Only a few spirochaetes can be cultured *in vitro* and their identification and diagnosis is based primarily on serological tests. The order Spirochaetales is divided into two families and five genera, of which three, *Treponema*, *Borrelia* and *Leptospira* cause human disease (Table 15.1).

TREPONEMA

Definition. Gram-negative spiral bacteria; motile; non-capsulate; non-sporing; non-culturable *in vitro*.

Classification. The treponemal species that cause human disease are *Treponema pallidum* and *T. carateum*. *T. pallidum* causes syphilis and a sub-species, referred to as *T. pertenue*, causes yaws. These organisms differ only in the clinical conditions they cause. *T. carateum* causes pinta.

T. pallidum

Epidemiology. *T. pallidum* is a strict human pathogen causing syphilis. Spread is primarily by sexual transmission; congenital infection can occur.

Morphology and identification. The organism is a thin, coiled spirochaete, 0.5 × 10.0 μm, which cannot be grown *in vitro* and grows only slowly *in vivo*. In clinical specimens, motile forms of the organism can be visualized by dark field microscopy or following specialized staining.

Associated infections
- Syphilis: this occurs worldwide and is the third most common sexually transmitted disease in the developed world. The organism enters the body

73

SPIROCHAETES	
Spirochaete	Associated infections
Treponema	
T. pallidum	Syphilis
T. pertenue	Yaws
T. carateum	Pinta
Borrelia	
B. recurrentis	Relapsing fever
B. burgdorferi	Lyme disease
Leptospira	Flu-like illness
	Meningitis
	Weil's disease

Table 15.1 Spirochaetes and associated infections.

via skin or mucous membrane abrasions. Localized multiplication occurs, resulting in inflammatory cell infiltration, followed by endarteritis. The infection is divided into three phases: primary, secondary and tertiary.
• Congenital syphilis: transmission to the fetus may present as intrauterine death or congenital abnormalities.

Laboratory diagnosis. As *T. pallidum* cannot be grown *in vitro* on standard laboratory media, diagnosis is based on microscopy and serology:
• *Microscopy.* Primary, secondary and congenital syphilis can be diagnosed by dark field examination of fresh material from skin lesions.
• *Serology.* Diagnosis of syphilis in most patients is based on serological tests. Two types of test are employed: non-specific and specific treponemal tests. The interpretation of serological tests for syphilis in relation to disease activity is discussed later.
❙ Non-specific antibody tests: these rapid tests determine the presence of IgM and IgG antibodies, which develop against lipids released from damaged *T. pallidum* during the early stages of the disease. The antigen used in these assays, 'cardiolipin', is extracted from beef heart; cardiolipin binds antibodies to *T. pallidum* lipids. The assays are therefore non-specific but are useful in the diagnosis of active syphilis. The tests commonly used are the venereal disease research laboratory (VDRL) test and the rapid plasma reagent (RPR) test. Both assays are based on the agglutination of cardiolipin antigen by the patient's serum on a glass slide. As the assays are non-specific, false-positive results are common, and occur in other conditions (e.g. leprosy, tuberculosis, viral infections, malaria, rheumatoid arthritis); results need confirmation by specific serological tests.

2 Specific antibody tests are based on *T. pallidum* antigens and are used to confirm non-specific screening tests. Commonly used tests include the fluorescent treponemal antibody (FTA) test and the *T. pallidum* haemagglutination assay (TPHA), in which erythrocytes coated with *T. pallidum* antigen are agglutinated by serum from patients with antibodies to *T. pallidum*. Positive results may reflect old disease as they remain positive after treatment. False-positive results may occur in other treponemal infections, e.g. yaws, pinta.

Specific tests to detect IgM antibodies are used to diagnose congenital injections.

T. carateum

Pinta is caused by *T. carateum* and is found primarily in Central and South America. It occurs in all age groups. After a 1–3 week incubation period, small papules develop on the skin. These can enlarge, last for several years and result in hypopigmented lesions. Spread is via direct contact or insect vectors. Diagnosis and treatment is as for syphilis.

T. pertenue

T. pertenue is the causative agent of yaws. This is primarily a skin disease; destructive lesions (granuloma) of the skin, lymph nodes and bone can occur. Yaws is found primarily in South America and Central Africa.

Spread is by direct contact with infected skin lesions. Diagnostic procedures and treatment are as for syphilis.

BORRELIA

Borrelia are associated with two important human infections, relapsing fever (primarily *B. recurrentis*) and Lyme disease (*B. burgdorferi*).

Definition. Gram-negative, spiral bacteria; motile; non-capsulate; non-sporing; require specialized media for culture; anaerobic or microaerophilic.

B. recurrentis, B. duttoni and other Borrelia

Relapsing fever is characterized by episodes of fever and septicaemia, separated by periods when the patient is apyrexial:
- epidemic relapsing fever is caused by *B. recurrentis* and is spread via the human body louse;

- endemic relapsing fever is caused by many species of *Borrelia* and is spread from rodents via infected ticks. The insect can infect their young trans-ovarily, maintaining an endemic reservoir.

The characteristic relapsing picture is related to the ability of *Borrelia* to vary their antigenic structure and avoid specific host antibodies.

B. recurrentis can be identified in stained blood smears, taken when a patient is pyrexial. Isolation may be successful with enriched media in anaerobic conditions. Serological tests are not often helpful due to antigen variation. Tetracycline is the treatment of choice.

B. burgdorferi

B. burgdorferi is the cause of Lyme disease, a recently described multisystem infection. The organism is transmitted to man by the *Ixodes* tick which is associated particularly with deer. Infection occurs mainly in forested areas and is characterized initially by a skin lesion at the site of the tick bite but spreads to involve joints, the heart and the central nervous system.

Diagnosis is by serological tests. Penicillin or tetracycline are effective antibiotics. Prevention is by avoidance of tick bites.

LEPTOSPIRA

Definition. Gram-negative, coiled bacteria; non-capsulate; non-sporing; motile; aerobes which grow slowly in enriched media.

Classification. The genus *Leptospira* has two species, *L. interrogans* and *L. biflexa*. *L. interrogans* causes disease in man, whereas *L. biflexa* is a free-living saprophyte. There are over 100 serotypes of *L. interrogans* but only a few are associated with human disease, namely *icterohaemorrhagiae*, *canicola* and *pomona*.

Epidemiology
- Leptospirosis occurs worldwide. Animals act as a reservoir of human infection, e.g. rats (*icterohaemorrhagiae*), dogs (*canicola*) and pigs (*pomona*).
- Human infections occur mainly following exposure to water contaminated with infected urine or from bites. At-risk groups include sewer workers, farmers and those involved in watersports.

Morphology and identification. Leptospires are motile organisms with two periplasmic flagellae, each anchored at opposite ends of the organism.

They can be grown on specially defined media enriched with serum, although incubation for up to 2 weeks may be required.

Associated infections. Infections range from a mild flu-like illness to meningitis and Weil's disease (with liver and renal damage).

Laboratory diagnosis. Blood or urine may be cultured in specialized media. However, diagnosis is often made by serology for the presence of leptospiral antibodies.

Treatment and prevention
- Treatment is with penicillin or tetracycline.
- Rodent control measures help to prevent infection. Groups at risk should cover cuts and abrasions.

CHAPTER 16

Mycobacteria

MYCOBACTERIA

Definition. Acid-fast bacilli; obligate aerobes; non-capsulate; non-motile; grow slowly on specialized media; cell walls have large lipid content.

Classification
- The genus includes: *Mycobacterium tuberculosis* and *M. bovis* (tuberculosis); *M. leprae* (leprosy); also the atypical mycobacteria, including *M. avium-intracellulare*, *M. kansasii* and *M. marinum*.
- Mycobacteria stain poorly by Gram-stain due to cell-wall mycolic acid. They are recognized by the Ziehl–Neelsen (ZN) stain (Plate 23).
- Mycobacteria grow only slowly; visible colonies appear after 1–12 weeks depending on species (Plate 24).
- Classification is based on culture characteristics, including nutritional requirements, rate of growth, pigmentation and biochemical properties.

M. tuberculosis and M. bovis

M. tuberculosis (primary host, man; commonly called the tubercle bacillus) and *M. bovis* (primary host, cattle) are very similar species of mycobacteria and cause similar infections.

Epidemiology
- *M. tuberculosis* infections are spread usually by inhalation of droplets; rarely by ingestion. Human infection with *M. bovis* is acquired via contaminated milk. Tuberculosis is highly infectious and outbreaks may occur.
- Mycobacteria are able to survive for long periods in the environment because they withstand drying.

Pathogenicity
- *M. tuberculosis*, in common with other mycobacteria, is an intracellular

pathogen; its survival within macrophages is related to its ability to prevent phagosome–lysosome fusion.
• The host response to mycobacterial infection is via the cell-mediated immune system and results in the formation of granulomas. *M. tuberculosis* may be dormant within granulomas; reactivation results in secondary or post-primary tuberculosis.

Associated infections
• Pulmonary tuberculosis.
• Extrapulmonary tuberculosis, including: meningitis; osteomyelitis; miliary tuberculosis (multisystem infection); cervical and mesenteric lymphadenopathy; abdominal tuberculosis; and renal tuberculosis.
• *M. bovis* infection is typically localized to bone marrow and cervical or mesenteric lymph nodes.

Laboratory diagnosis
• Microscopy of smears of sputum, urine or tissue by ZN stain; stained bacilli appear as thin bacilli with beads. A fluorescent rhodamine-auramine dye can also be used.
• Culture on special media for up to 12 weeks, e.g. Lowenstein–Jensen medium, which contains egg yolk, glycerol and mineral acids plus inhibitors, such as malachite green, to reduce growth of other bacteria. Specimens, e.g. sputum, contaminated with normal flora, are pre-treated with acid to reduce contamination.
• Immunological tests to detect *M. tuberculosis* (e.g. Mantoux test or Heaf test): these are based on the inoculation of purified protein derivative (PPD), derived from tubercle culture filtrate, into the patient's skin to demonstrate cell-mediated immunity to *M. tuberculosis*. In some countries, the widespread use of vaccination, or the frequency of previous exposure to tuberculosis, decreases the usefulness of skin testing.

Treatment and prevention
• Treatment is with combinations of antimycobacterial drugs (e.g. rifampicin, isoniazid and pyrazinamide) for prolonged periods ($>$ 6 months). Multiply-resistant strains of *M. tuberculosis* are, however, emerging.
• For prevention of *M. tuberculosis* infection different strategies have been tried, including widespread immunization with a live attenuated vaccine (Bacille Calmette–Guérin (BCG)) and isolation and prompt treatment of cases combined with chemoprophylaxis of close contacts.
• *M. bovis* infection is prevented by destruction of tuberculin-positive cattle and milk pasteurization.

M. leprae

M. leprae (Hansen's bacillus) causes leprosy.

Epidemiology
- There are over 10 million cases of leprosy worldwide, mainly in Asia, Africa, and South America. There are no animal reservoirs.
- Transmission follows prolonged exposure to shedders of the bacilli via respiratory secretions. The incubation period is 2–10 years; without prophylaxis up to 10% develop the disease.

Morphology and identification. *M. leprae* is an acid-fast bacillus. It can be grown in the footpads of mice or armadillos, from tissue culture and on artificial media.

Associated infections. There are two major types of leprosy; lepromatous and tuberculoid, with various intermediate stages.
- Lepromatous: progressive infection, resulting in nodulous skin lesions and nerve involvement; associated with a poor prognosis.
- Tuberculoid: a more benign, non-progressive form which involves macular skin lesions and severe asymptomatic nerve involvement. Spontaneous healing usually results following tissue and nerve destruction.

Laboratory diagnosis. Microscopy of scrapings from skin or nasal mucosa, or skin biopsies, examined after Ziehl–Neelsen staining.

Treatment and control
- Dapsone: used in treatment but resistance is increasing and rifampicin is often added.
- Chemoprophylaxis is needed for close contacts of contagious individuals; this is particularly important for young children in contact with adults with leprosy.

Atypical mycobacteria

These grow at a variety of temperatures; some are rapid growing (3–4 days), others are slower (>8 weeks). Some produce pigmented colonies in light (photochromogens) or in light and dark (scotochromogens).

They are transmitted to man primarily from environmental or animal sources. Associated infections are listed in Table 16.1. Some species, e.g. *M. avium-intracellulare*, are resistant to first-line anti-mycobacterial agents.

ATYPICAL MYCOBACTERIA

Mycobacteria species	Associated infections
M. ulcerans	Skin ulcers (tropics)
M. kansasii	Chest disease
M. marinum	Granulomatous skin lesions associated with cleaning fish tanks and swimming pools
M. chelonei	Cutaneous abscesses; occasionally disseminated infection in the immunocompromised
M. avium-intracellulare	Cervical lymphadenopathy in children; chest infection and invasive infections in AIDS patients

Table 16.1 Infections associated with atypical mycobacteria.

CHAPTER 17

Chlamydiae, Rickettsiaceae, Mycoplasma and L-forms

CHLAMYDIAE

Definition. Obligate intracellular bacteria; normally grown in tissue cultures.

Classification. The genus *Chlamydia* is divided into three species: *C. psittaci*; *C. trachomatis* and *C. pneumoniae*.

Classification is based on antigens, morphology of intracellular inclusions and disease patterns (Table 17.1).

Morphology and identification. All chlamydiae share a common group antigen but may be distinguished by species-specific antigens. They multiply in the cytoplasm of host cells following a well-defined developmental cycle.

Chlamydiae have two distinct morphological forms:
1 the elementary body (EB): an infectious extracellular particle (300–400 nm);
2 the reticulate body (RB): an intracellular non-infectious particle (800–1000 nm).

Life cycle
- EBs bind to specific host cell receptors and enter by endocytosis. Target cells include conjunctival, urethral, rectal and endocervical epithelial cells.
- Intracellular EBs remain within phagosomes and replicate. Lysosomal fusion with the phagosome-containing EBs is inhibited, probably by chlamydial cell-wall components.
- After approximately 8 h, EBs become metabolically active and form RBs. RBs synthesize DNA, RNA and protein utilizing host cell adenosine triphosphate (ATP) as an energy source.
- RBs undergo multiple division and the phagosome becomes an 'inclusion body' (visible by light microscopy).
- After approximately 24 h, the RBs reorganize into smaller EBs and, after a further 24–48 h, the host cell lyses and infective EBs are released.

CHLAMYDIA			
Species	Serotypes	Natural host	Common infections
C. trachomatis	A–C	Humans	Trachoma
	D–K	Humans	Cervicitis
			Conjunctivitis
			Urethritis
			Proctitis
			Pneumonia
	L1, L2, L3	Humans	Lymphogranuloma venereum
C. psittaci	1	Birds and some mammals	Pneumonia
C. pneumoniae	1	Humans	Acute respiratory infections

Table 17.1 Classification and common infections of Chlamydia.

Pathogenicity. Chlamydiae are intracellular parasites avoiding cellular defences by inhibiting phagosome–lysosome fusion.

C. trachomatis

Classification. C. trachomatis can be divided into different serotypes ('serovars'); serotypes A–C, D–K and L1, L2, L3.

Epidemiology. C. trachomatis serotypes D–K are found worldwide, whereas the LGV serotypes occur primarily in Africa, Asia and South America. The majority of C. trachomatis infections are genital and are sexually acquired. Asymptomatic infections can occur in women. Eye infections in adults probably result from oculo-genital contact. Similarly, neonates can be infected during birth from an infected mother.

Pathogenesis. C. trachomatis gains access via disrupted mucous membranes and can infect mucosal columnar or transitional epithelial cells resulting in severe inflammation. In lymphogranuloma venereum (LGV), the lymph nodes, which drain the site of the primary infection, are involved.

Associated infections
• Trachoma: keratoconjunctivitis caused by serotypes A–C; the leading cause of preventable blindness in the world.

- Adult inclusion conjunctivitis: a milder disease than trachoma; caused by serotypes D–K; associated with genital infections.
- Neonatal conjunctivitis: *C. trachomatis* can be acquired at birth from an infected mother; associated with serotypes D–K.
- Neonatal pneumonitis: presents 3–12 weeks after birth; often associated with neonatal conjunctivitis.
- LGV: a sexually transmitted disease, with genital ulcers and suppurative inguinal adenitis. It is common in tropical climates and associated with serotypes L1–L3.
- Urogenital infections: associated with serotypes D–K; the most common cause of sexually transmitted disease in the developed world. Complications include pelvic inflammatory disease and perihepatitis in women, and epididymitis and Reiter's syndrome in men.

Laboratory diagnosis
- Culture: swabs from the affected site collected in a special transport medium are inoculated into tissue culture. Growth of *C. trachomatis* is recognized by Giemsa staining to show intracellular bodies or by immunofluorescence staining with specific antisera.
- Direct antigen detection can be made by immunofluorescence (Plate 25) or enzyme-linked immunosorbent assay (ELISA) methods. Both techniques utilize labelled specific antibodies to *C. trachomatis*.
- Serology: antibodies to *C. trachomatis* may be detected by complement fixation or micro-immunofluorescence tests. Serological testing has limited value as it is difficult to distinguish between past and current infection. Assays based on the polymerase chain reaction and ligase chain reaction are also now available.

C. psittaci

C. psittaci causes infection in psittacine birds (e.g. parrots) and may be transmitted to humans.

Epidemiology. Transmission to humans is via inhalation, usually of dried bird guano. Pet shop workers, bird fanciers and poultry farm workers are at particular risk of developing psittacosis.

Associated infections. Flu-like illness and pneumonia.

Laboratory diagnosis. By serology, e.g. complement fixation test. Culture of *C. psittaci* is also possible but is only performed in laboratories with high-level containment facilities.

Treatment. Tetracycline is the drug of choice for treating *C. psittaci* infections.

C. pneumoniae

Originally designated TWAR (Taiwan-associated respiratory disease) agent, but now called *C. pneumoniae* and is an established cause of atypical pneumonia.

Infection is spread via direct human contact, with no apparent animal reservoir. It is associated with respiratory infections, including pharyngitis, sinusitis and pneumonia (which can be severe, particularly in the elderly). Diagnosis is by serology and treatment includes either tetracycline or erythromycin.

RICKETTSIACEAE

Rickettsiaceae cause a number of important human infections including typhus, the related spotted fevers, and Q-fever.

Definition. Small, Gram-negative bacilli (0.2 × 0.7 μm diameter); obligate intracellular pathogens; utilize ATP from host cell; grow only in tissue culture.

Classification. Rickettsiaceae contains two important genera, *Rickettsia* and *Coxiella*.

Rickettsia

The genus contains a number of species that cause human infection (Table 17.2).

Epidemiology. Rickettsial infections are zoonoses with a variety of animal reservoirs and insect vectors.

Associated infections. Typhus and spotted fevers.

Laboratory diagnosis. This is by serology. The Weil–Felix test detects cross-reacting antibodies to Rickettsiae which agglutinate certain strains of *Proteus* (OX-19 and OX-2). The test is non-specific and has now been superseded by more specific serological tests based on purified rickettsial antigens (immunofluorescence assays and ELISA).

RICKETTSIA

Species	Principal host/ reservoir	Vector	Disease
R. typhi	Rats	Fleas	Murine typhus
R. prowazeki	Man/squirrels	Lice	Louse-bound typhus
R. tsutsugamushi	Rats	Mites	Scrub typhus
R. akari	Mice	Mites	Rickettsial pox
R. rickettsii*	Dogs	Ticks	Rocky mountain spotted fever

* Tick-borne spotted fevers in areas of the world outside the USA carry a variety of names and are caused by Rickettsiae very similar to R. rickettsii (e.g. Boutonneuse fever caused by R. conorrii occurs in the Mediterranean region).

Table 17.2 Vectors and infections associated with Rickettsia.

Treatment and prevention. Treatment is with tetracycline and chloramphenicol. Infection can be prevented by avoidance of the various vectors. A vaccine to R. prowakzekii is available.

Coxiella

This is an animal pathogen that occasionally infects man via tick bites or inhalation of dried, contaminated faecal matter. At-risk groups include abattoir workers. Milk may occasionally act as a vector.

Coxiella burnetti causes Q-fever and endocarditis. Diagnosis is by serology (complement fixation and immunofluorescence tests). Treatment is with tetracycline.

MYCOPLASMA AND UREAPLASMA

Definition. Small (<1.0 μm), Gram-negative organisms; pleomorphic with no cell wall; grow slowly on enriched media; aerobic or facultatively anaerobic.

Classification. The family Mycoplasmataceae consists of two genera, *Mycoplasma*, with 69 species and *Ureaplasma* with two species. Only three species have been identified as human pathogens: *Mycoplasma pneumoniae*, *M. hominis* and *Ureaplasma urealyticum*.

Epidemiology. M. pneumoniae infection is common worldwide and may cause up to 30% of acute lower respiratory tract infections; epidemics can occur. Infection is spread by nasal secretions. *M. hominis* and *U. urealyticum* colonize the genitourinary tracts.

Associated infections
- *M. pneumoniae*: pharyngitis; community-acquired pneumonia.
- *U. urealyticum/M. hominis*: urethritis; possible cause of pelvic inflammatory disease and opportunist infection in the immunocompromised (arthritis, meningitis).

Laboratory diagnosis
- Culture: mycoplasmas and ureaplasmas can be grown on enriched media; penicillin is often added to inhibit other organisms. Although *M. pneumoniae* can be isolated from sputum following incubation for up to 3 weeks, diagnosis is normally by serology. *M. hominis* grows after approximately 4 days, producing colonies with a 'fried egg' appearance. *U. urealyticum* requires urea for growth and forms small colonies.
- Serology: *M. pneumoniae* infections can be diagnosed by serological tests for IgG (fourfold rise in titres is indicative of current infection) or IgM.

Treatment. This is with macrolides (e.g. erythromycin) or tetracyline.

L-FORMS

L-forms, named after the Lister Institute where they were first reported, are cell-wall-deficient forms of bacteria. Removal of the cell wall can be achieved by cell-wall-acting antibiotics. L-forms need to be maintained on osmotically stabilized media to prevent lysis. L-forms are able to multiply and their colonial morphology is similar to the 'fried egg' appearance of the mycoplasmas. A few L-forms are also able to reform their cell walls and revert back to parent vegetative state with a complete cell wall. These are referred to as unstable L-forms. Other L-forms cannot produce new cell walls and are referred to as stable L-forms. It has been suggested that mycoplasmas are stable L-forms.

Protoplasts are forms of bacteria which have also had their cell walls removed. They are metabolically active and can grow, but are unable to multiply. They only survive in an osmotically stabilized medium.

Spheroplasts are protoplasts but with some cell wall remaining.

L-forms have been associated with renal infections but their clinical significance is unclear.

CHAPTER 18

Basic Virology

VIRUS STRUCTURE (Fig. 18.1)

Size and shape. Viruses range in size from 20 to 300 nm in diameter and many have typical shapes that aid identification.

Genome. Viruses contain either DNA or RNA. Nucleic acids represent the main component of the virus core and are associated with core proteins. Viral nucleic acids range from 1.5×10^6 Da (parvoviruses) to 200×10^6 Da (poxviruses) and may be single-stranded or double-stranded, circular or linear, a single molecule or in segments.

Capsid. A protein coat enclosing the genome and core proteins and consisting of capsomeres (capsid subunits).

Envelope. Lipid bilayer membrane surrounding the nucleocapsid of some viruses (enveloped viruses); the envelope carries glycoproteins, which form projections or spikes.

Virion. The complete infectious virus particle; may lack nucleic acids (empty particles) or carry defective genomes (defective particles), which can interfere with normal replication (defective interfering particles).

VIRUS CULTIVATION

Viruses are obligate, intracellular parasites and thus can only replicate in living cells. Viruses utilize the host cell metabolism to assist in the synthesis of viral proteins and progeny virions; the host cell range of viruses may be narrow or wide.

For diagnostic purposes, most viruses are grown in cell cultures, either secondary or continuous cell lines; the use of embryonic eggs and laboratory animals for virus culture is reserved for specialized investigations. Cell culture lines for virus cultivation are often derived from tumour tissues, which can be passaged indefinitely.

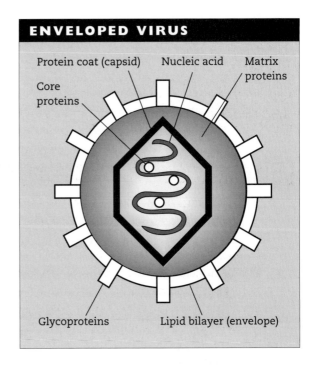

Fig. 18.1 An enveloped virus.

Virus replication in cell cultures may be detected by:
- *cytopathic effect (CPE)*: some viruses can be recognized by their effect on cell architecture, e.g. necrosis, lysis, the presence of inclusion bodies, or the formation of multinucleated cells;
- haemadsorption: viruses expressing haemagglutinins on the cell surface may be recognized by absorption of red blood cells to infected cells;
- immunofluorescence: the appearance of virus-coded proteins on the surface, in the nucleus or cytoplasm of infected cells may be detected by immunofluorescence techniques using virus protein-specific antibody.

VIRAL INFECTION OF HOST CELLS

Virus replication in host cells involves the following steps.

1 Attachment (adsorption) occurs of the viral nucleocapsid (naked viruses) or of virus envelope components (enveloped viruses) to cell surface molecules (receptors); this involves specific interaction between viral glycoproteins (e.g. the haemagglutinin of influenza virus) and host cell surface components (e.g. *N*-acetyl neuraminic acid for influenza virus). Many viruses have highly specific receptors.

2 Penetration of the virus into the host cell takes place (often by receptor-mediated endocytosis).

3 Uncoating follows, which involves the proteolytic removal of viral protein coat and liberation of nucleic acid and attached core proteins.

4 In order to direct the host cell ribosome to produce viral proteins (core, capsid), virus-specific mRNA must be produced. The mechanisms for virus-specific mRNA production depend on the viral genome type:

    (a) RNA or DNA;

    (b) single-stranded or double-stranded;

    (c) positive sense (base sequence configured as required for translation = mRNA sense) or negative sense (base sequence requires transcription).

Examples are shown in Fig. 18.2.

The mechanisms for replication of viral nucleic acid also depend on viral genome type; examples are given in Fig. 18.3.

5 Morphogenesis and maturation occur with assembly of components (nucleic acid, proteins) to form subviral particles (pre-core, core particles) and viral particles (virions, empty particles, DI particles).

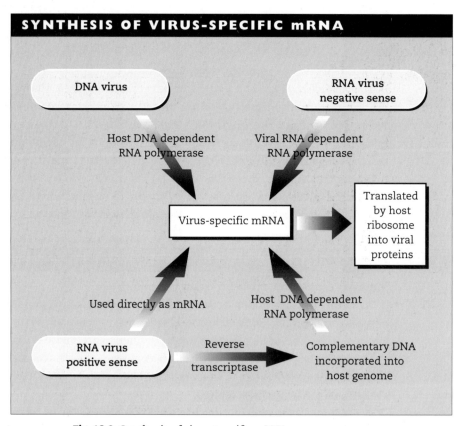

Fig. 18.2 Synthesis of virus-specific mRNA.

Viral infection of host cells

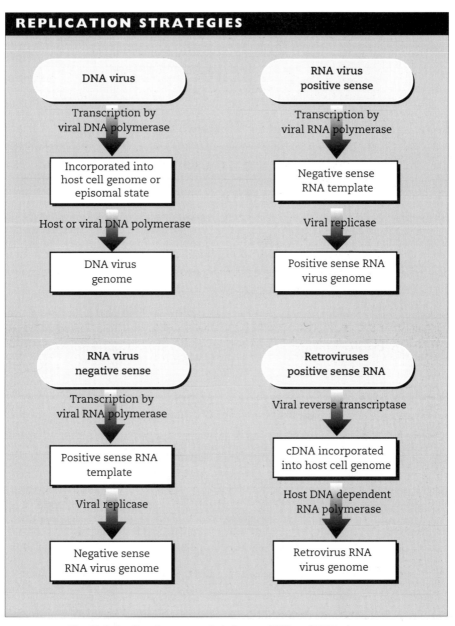

Fig. 18.3 Replication strategies of some DNA and RNA viruses.

6 Release of the virus is by bursting of infected cells (lysis) or by budding through plasma membrane (host cell does not necessarily lose viability, therefore it can shed viral particles for extended periods).

With some viruses, e.g. hepatitis B, the host cell remains viable and

VIRAL CLASSIFICATION

Organ system involved (disease)	Examples of clinically important viruses
Systemic infections	Measles virus, rubella virus, chickenpox virus, enteroviruses, retroviruses
Central nervous system	Polio- and other enteroviruses; rabies virus, arthropod-borne viruses, herpes simplex virus, measles virus, mumps virus, retroviruses, cytomegalovirus
Respiratory tract (common cold, tracheitis, bronchitis, bronchiolitis, pneumonia)	Influenzavirus, parainfluenzavirus, respiratory syncytial virus, adenovirus (enteroviruses), cytomegalovirus
Eye (conjunctivitis, retinitis)	Herpes simplex virus, adenovirus, cytomegalovirus
Skin and mucous membranes (rash, warts)	Herpes simplex virus, papillomavirus, chickenpox virus, measles virus, rubella virus, parvovirus
Liver	Hepatitis viruses A–G, yellow fever virus, herpes viruses, rubella virus
Salivary glands	Mumps virus, cytomegalovirus
Gastrointestinal tract	Rotaviruses, adenoviruses, astroviruses, caliciviruses

Note: this classification is clinically orientated and not virus systematic. The same clinical symptoms can be caused by numerous viruses, and one virus can cause different clinical syndromes.

Table 18.1 Viral classification according to disease or organ system involved.

continues to release virus particles or subviral antigens at a slow rate. These persistent infections act as a continuing source of new infectious viruses.

During latent infections the virus does not undergo replication; the viral nucleic acid may remain in host-cell cytoplasm (e.g. herpes virus) or become incorporated into the host genome (e.g. human immunodeficiency virus (HIV)). A trigger is required to recommence viral nucleic acid replication, transcription and translation.

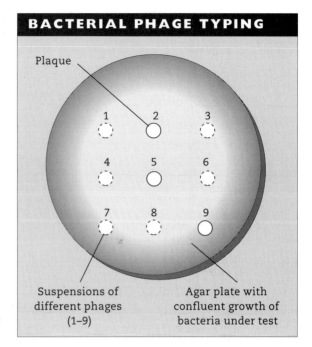

Fig. 18.4 Bacterial phage typing. The agar plate is inoculated with bacteria. Bacteriophages (1–9) are spotted onto the plate. After overnight incubation bacterial lysis is indicated by areas of no bacterial growth. The bacteria is susceptible to phages 2, 5 and 9.

BACTERIOPHAGES

Bacteriophages are viruses that infect bacteria. They are important in medical microbiology for a number of reasons (see p. 16).

Phage typing. Bacteriophages initiate infection by attachment to the bacterial surface; this is specific and phages have a very limited host range, even within a bacterial species. For example, different isolates (strains) of *Staphylococcus aureus* may be susceptible to different bacteriophages, i.e. they show different phage typing patterns. Susceptibility to a bacteriophage is recognized in the laboratory by observing bacterial lysis (Fig. 18.4).

Transduction. Transfer of genetic information takes place from one bacterial cell to another mostly through naked plasmids.

Bacterial pathogenicity. Some bacteriophages code for toxins, which, when translated in the host bacteria, result in toxin production and clinical disease. The toxins of *C. diphtheriae* and Group A streptococci are encoded on bacteriophages.

CLASSIFICATION OF VIRUS FAMILIES

Nucleic acid type	Symmetry of nucleocapsid	Envelope	Virus family
DNA SS Linear	Icosahedral	−	*Parvoviridae*
DNA DS Circular	Icosahedral	−	*Papovaviridae*
DNA DS Linear	Icosahedral	−	*Adenoviridae*
DNA DS Linear	Icosahedral	+	*Herpesviridae*
DNA DS Linear	Complex	+	*Poxviridae*
DNA DS Circular	Icosahedral	+	*Hepadnaviridae*
RNA SS Linear	Icosahedral	−	*Picornaviridae*
RNA SS Linear	Icosahedral	−	*Caliciviridae*
RNA SS Linear	Icosahedral	+	*Togaviridae*
RNA DS	Icosahedral	−	*Reoviridae*
RNA SS	Complex	+	*Flaviviridae*
RNA SS	Complex	+	*Arenaviridae*
RNA SS	Icosahedral	+	*Coronaviridae*
RNA SS	Icosahedral	+	*Retroviridae*
RNA SS	Helical	+	*Orthomyxoviridae*
RNA SS	Helical	+	*Bunyaviridae*
RNA SS	Helical	+	*Paramyxoviridae*
RNA SS	Helical	+	*Rhabdoviridiae*

DS, double-stranded; SS, single-stranded; −, absent; +, present.

Table 18.2 Classification of virus families according to nucleic acid and virion structure.

VIRUS CLASSIFICATION

Viruses can be classified according to:
1 disease or organ system involved (Table 18.1);
2 nucleic acid type/virion structure (virus families; Table 18.2);
3 replication strategy (see Fig. 18.3).

DETECTION OF VIRUS INFECTIONS

1 Direct (detection of virus particles, viral antigen, viral lesions or nucleic acid) by: microscopy; electron microscopy; particle agglutination; immunoflourence; serology; PCR and RT-PCR.
2 Indirect (detection of virus-specific host response) by serology (complement fixation test, haemagglutination inhibition test, ELISA etc).

CHAPTER 19

Major Virus Groups

ENTEROVIRUSES

The enteroviruses (a genus of the picornaviridae) cause a variety of diseases including paralytic poliomyelitis, pleurodynia, myocarditis, meningitis, encephalitis and respiratory illness.

This genus of viruses includes:
- polioviruses (types 1, 2, 3);
- coxsackie A viruses (types 1–24);
- coxsackie B viruses (types 1–6);
- echoviruses (types 1–34);
- enteroviruses (types 68–71). (Enterovirus 72 now reclassified as hepatitis A virus, HAV).

Viruses. Icosahedral, non-enveloped RNA viruses, 28–30 nm in diameter.

Diagnosis
- Viral growth in monkey kidney cells from faeces, throat swabs or cerebrospinal fluid (CSF).
- Serology is by enterovirus IgM-specific ELISA or neutralization tests.

Pathogenesis. The incubation period is 7–14 days; primary replication takes place in the oropharynx and intestine, followed by viraemia. Specific damage in CNS (anterior horn cells of spinal cord) or muscle tissue (heart, skeletal muscle). Long-lasting local IgA and humoral IgM/IgG response develops after natural infection.

Epidemiology. Enteroviruses have worldwide distribution. Infections occur in infancy in developing countries and in early childhood in developed countries. Diseases (Table 19.1) are more severe in the elderly.

Prevention and control. Both live and killed poliovirus vaccines are available.

RHINOVIRUSES

Viruses. RNA viruses similar to enteroviruses (both are genera of picornaviridae). There are more than 100 serotypes. The rhinoviruses are the most common cause of 'common cold'.

Diagnosis. Specimens are isolated from nasal swabs in cell culture (HeLa cells).

Pathogenesis. The incubation period is 2–4 days. Replication takes place in the nasal mucosa leading to local IgA response.

Associated disease. Common cold.

Treatment. Symptomatic. There have been recent trials of treatment with alpha-interferon.

Epidemiology. Rhinoviruses are endemic worldwide. Transmission is by droplets or close contact.

CORONAVIRUSES

The coronaviruses are a frequent cause of the 'common cold'.

Virus. Enveloped RNA viruses.

Diagnosis. This is determined by: growth in tissue culture; immune electronmicroscopy; serology (enzyme-linked immunosorbent assay (ELISA)).

Pathogenesis. The incubation period is 2–5 days. Replication takes place in the epithelial cells of the respiratory tract and gut with development of strain-specific antibody. Asymptomatic shedding in faeces is common.

Associated infection. Common cold.

Epidemiology. This is a common infection; 90% of adults in the UK have specific antibody and during seasons of respiratory illness there is a 15% incidence.

Control. No specific vaccine.

INFLUENZAVIRUSES

The influenzaviruses are a common cause of respiratory tract infection.

ENTEROVIRUSES: ASSOCIATED DISEASES

Virus	Disease
Polio virus	Abortive and paralytic poliomyelitis or aseptic meningitis
Coxsackie virus	
Echovirus	
Coxsackie A virus	Herpangina; hand, foot and mouth disease
Coxsackie B virus	Pleurodynia, epidemic myalgia, myocardiopathy

Table 19.1 Enteroviruses: associated diseases.

Virus. Enveloped RNA virus of 100 nm diameter; there are three types A, B and C, determined by nucleoprotein. The envelope contains two glycoproteins in lipid bilayer: haemagglutinin (HA) and neuraminidase (NA). Variations in HA and NA determine subtypes (in humans, H1, H2, H3, N1, N2). Nomenclature is based on type, origin, strain, year of isolation and subtype e.g.

Type	Host	Origin	Strain number	Year	Subtype	=
A	Human	Puerto Rico	8	1934	H1N1	A/PR/8/34 (H1N1)

Diagnosis. This is by virus isolation in tissue culture (monkey kidney cells); immunofluorescence of nasopharyngeal aspirates; serology (complement fixation test; haemagglutination inhibition test).

Pathogenesis and immune response. Spread by droplets; infection of upper respiratory tract; incubation period of 1–4 days followed by virus replication in upper respiratory tract; antibody in serum (IgM) and secretions (IgA) appears about day 6. HA is an important virulence determinant.

Associated disease. Influenza.

Epidemiology. Antigenic variation is an important factor in global epidemiology.
- Antigenic 'drift': minor antigenic changes occur in HA and NA (within subtype) due to sequential point mutations (types A, B and C), leading to epidemics.
- Antigenic 'shift': major antigenic changes occur in HA and NA (change in subtype) due to genetic reassortment (type A only), leading to pandemics.

Treatment. Amantadine or rimantadine, for prophylaxis only.

Prevention. Inactivated vaccines (whole virus or subunit given intramuscularly) have an efficacy of 60–80% provided vaccine components and current wild-type viruses are sufficiently similar.

PARAMYXOVIRUSES

Viruses. Spherical or pleomorphic enveloped RNA viruses, 150–300 nm in diameter (the envelope contains HA, NA or other glycoproteins).

Classification. There are four genera:
1. paramyxoviruses (parainfluenzaviruses types 1–4);
2. rubulavirus (mumps virus);
3. morbillivirus (measles virus);
4. pneumovirus (respiratory syncytial virus (RSV)).

Diagnosis. This is made from: virus isolation in cell culture (monkey kidney cells); direct immunofluorescence tests; serology by various methods.

Pathogenesis and immune response. Transmission is by droplets, aerosols or direct person-to-person contact, followed by replication in the respiratory tract and viraemia. Local IgA response; IgM and IgG response leads to immunity for life against mumps and measles virus, not against RSV.

Associated diseases (see Table 19.2)

Epidemiology
- Parainfluenzaviruses and RSV are widely distributed geographically. Spread is by person-to-person contact or droplet.

PARAMYXOVIRUSES: ASSOCIATED DISEASES	
Virus	Disease
Parainfluenza virus	Tracheitis, croup, bronchiolitis, pneumonia
Respiratory syncytial virus (RSV)	Bronchiolitis in infants and young children
Mumps virus	Parotitis, pancreatitis, orchitis, meningitis
Measles virus	Measles

Table 19.2 Paramyxoviruses: associated diseases.

- Mumps is endemic worldwide. Cases occur throughout the year. The highest incidence is among 5- to 15-year-old children. Interfamily infections are common.
- Measles virus transmission by the respiratory route is widespread. Infections depend on the state of immunity and the size of the population. Rapid epidemics occur when the virus is introduced into isolated, susceptible communities.

Prevention and control
- RSV disease has an immunopathological component. Experimental inactivated vaccine has proved unsuccessful.
- Mumps: live attenuated virus vaccine is available as monovalent form (mumps only), or in combination with rubella and measles vaccine (MMR). There has been a sharp decline in incidence and prevalence in the USA after 20 years of mumps vaccination.
- Measles: live attenuated virus vaccine is used in mass immunization programmes. The vaccine is >95% effective. The first vaccination is given at the age of 15 months (to avoid failure due to the presence of maternal measles antibody); earlier application is given in developing countries, as there is a high measles mortality rate in children <1 year old.

RUBELLA VIRUS

Virus. Enveloped RNA virus of 60 nm diameter.

Diagnosis
- Acute infection: specific IgM antibody is determined by ELISA. Virus isolation from the throat if sample is taken from one week before to one week after onset of the rash.

- The immune status is determined by detection of specific IgG antibody by single radial haemolysis test, latex agglutination test or ELISA.

Pathogenesis. The incubation period is 14 days, with primary replication in the cervical lymph nodes, the rash coinciding with appearance of serum antibody. Lifelong immunity follows natural infection.

Primary infection during the first 12 weeks of pregnancy often leads to generalized fetal infection leading to cataract, nerve deafness, cardiac abnormalities, hepatosplenomegaly, purpura, jaundice (=congenital rubella syndrome).

Associated disease. Rubella: rash and lymphadenopathy with transient arthralgia. Infection without a rash is common. Congenital rubella syndrome.

Epidemiology. This is a frequent childhood infection, transmitted via droplets. An epidemic occurs every 6–9 years. However, markedly reduced since vaccination programmes in place.

Treatment. Symptomatic. Proven infection during the first 3 months of pregnancy is an indication for therapeutic abortion.

Prevention and control. Live attenuated vaccine is used. Immunization programmes exist in most developed countries.

HERPESVIRUSES

The herpesviruses are a family of enveloped DNA viruses that characteristically cause latent infections. Following primary infection, viral DNA lies dormant in various tissues and may be reactivated. Herpesviruses include:
- herpes simplex virus (HSV; type 1 and type 2);
- varicella-zoster virus;
- cytomegalovirus;
- Epstein–Barr virus;
- human herpesviruses; types 6–8.

Herpes simplex virus (HSV)

Virus. Enveloped DNA virus, 120 nm in diameter. There are two virus types, HSV 1 and HSV 2. They can be distinguished serologically.

Diagnosis. This is made by: electronmicroscopy of vesicle fluid; direct fluorescent antibody staining of vesicle material; isolation in tissue culture (wide variety of cell lines); serology (e.g. ELISA); nucleic acid detection.

Pathogenesis. Spread is by direct contact. The incubation period is variable (1–30 days). There are local IgA, IgM and IgG responses, but virus avoids elimination and its DNA becomes dormant in sensory ganglia. Reactivation is associated with various stimuli (e.g. stress, immunocompromised status, bacterial infection).

Associated diseases
- Primary diseases are: stomatitis (HSV 1; less commonly HSV 2); herpes genitalis (HSV 2; less commonly HSV 1); keratoconjunctivitis.
- Recurrent diseases are: herpes labialis (cold sores); recurrent herpes genitalis; keratoconjunctivitis.
- Invasive infections can lead to: encephalitis; neonatal HSV infection (e.g. encephalitis); disease in the immunocompromised (e.g. pneumonitis, oesophagitis, encephalitis).

Treatment. Acyclovir.

Epidemiology. HSV has worldwide distribution. Ninety percent of adults show serological evidence of past infection.

Varicella zoster virus (VZV)

Virus. Enveloped DNA virus, 180 nm in diameter.

Diagnosis. This is made by: electronmicroscopy of vesicle fluid; direct fluorescent antibody staining of vesicle material; isolation in tissue culture (wide variety of cell lines); serology, nucleic acid detection.

Pathogenesis. Spread is by droplets or direct contact with vesicle fluid. The incubation period is 13–21 days. The virus enters via the respiratory tract followed by viraemia and a generalized rash (chickenpox). It may become latent in the sensory ganglia; a variety of stimuli (e.g. immunosuppression) result in reactivation with vesicles in various dermatomes (shingles).

Associated diseases. These include: chickenpox; shingles; disseminated VZV infection in the immunocompromised and neonates.

Treatment. Acyclovir in the early stages; zoster immune globulin (ZIG).

Epidemiology. VZV has a worldwide distribution.

Cytomegalovirus (CMV)

Virus. Enveloped DNA virus, 200 nm in diameter.

Diagnosis. This is made by: isolation in fibroblast cell lines; serology (IgM for active disease; IgG for immune status); tissue biopsy (typical 'owl's eye' inclusion bodies); nucleic acid detection methods (e.g. polymerase chain reaction (PCR)).

Pathogenesis. Spread is by contact with infectious secretions (e.g. oropharyngeal); the virus may also be acquired by blood transfusion or organ transplantation. The primary infection is often mild. CMV remains latent in various tissues and may cause secondary infections, particularly in the immunocompromised.

Associated diseases. Glandular fever syndrome. Disseminated infection occurs in the immunocompromised (e.g. transplant patients, acquired immunodeficiency syndrome (AIDS) patients), leading to pneumonitis, hepatitis and retinitis. Congenital infection can occur.

Treatment. Ganciclovir; CMV hyperimmunoglobulin.

Epidemiology. CMV has a worldwide distribution. Of adults, 60–90% show serological evidence of exposure to CMV.

Epstein–Barr virus (EBV)

EBV is the principal cause of glandular fever syndrome.

Virus. Enveloped DNA virus, 120 nm in diameter.

Diagnosis. This is by: culture (which is rarely performed); serology — test for heterophile antibodies (Paul–Bunnell test; antibodies that appear early in EBV infection and cross-react with sheep or horse erythrocytes, resulting in agglutination); IFT or specific ELISA for antibody to EBV capsid and nuclear antigens.

Pathogenesis. Spread is by contact with saliva; the primary site of infection is the epithelium of the pharynx. B-lymphocytes become infected, and EBV may lie dormant in these cells. Atypical monocytes in the blood ('infectious mononucleosis') are EBV-infected B-lymphocytes.

Associated diseases. Glandular fever syndrome ('infectious mononucleosis'). EBV is implicated in Burkitt's lymphoma and nasopharyngeal carcinoma. Glandular fever may be complicated by widespread EBV infection in a few patients (e.g. hepatitis, myocarditis, meningoencephalitis).

Treatment. No effective antiviral treatment has yet been developed.

Epidemiology. EBV has a worldwide distribution. By the age of 7 years old more than 50% of children have antibodies.

Human herpesvirus 6

Human herpesvirus 6 (HHV6) is a recently discovered member of the herpesvirus family. It is an enveloped DNA virus, 180 nm in diameter.

This is a cause of roseola infantum, also called exanthema subitum or 'sixth disease', which has an incubation period of 5–15 days, followed by high fever and maculopapular rash.

HHV7 is possibly a variant of HHV6.

HHV8 has been associated with the development of Kaposi's sarcoma.

PARVOVIRUSES

Virus. Small, round, featureless DNA viruses of 20 nm diameter; single serotype: human parvovirus (HPV) B19.

Diagnosis. This is made by serology: HPV-specific IgM by ELISA.

Pathogenesis. The incubation period is up to 17 days with replication in rapidly dividing cells, mainly erythroid progenitor cells.

Associated diseases. Erythema infectiosum or fifth disease: fever, chills, myalgia one week after infection, followed by maculopapular rash at 17 days. Aplastic crisis can occur in patients with chronic haemolytic anaemias (e.g. thalassaemia, spherocytosis). Hydrops fetalis and spontaneous abortions may result from fetal infection.

Epidemiology. It is a frequent childhood infection worldwide and is transmitted via the respiratory tract (droplets).

ROTAVIRUSES

Virus. RNA virus, 75 nm in diameter with wheel-like structure. Five groups (A–E), and within group A at least two subgroups (I, II) and various serotypes (14 VP7-specific or G types; >20 VP4-specific or P types; G1 to G4 viruses cause over 90% of human infections.).

Diagnosis. This is made by direct detection in faeces by electronmicroscopy, ELISA or latex agglutination test.

Pathogenesis and immune response. Infection is via the faecal–oral route with an incubation period of 1–2 days. Virus replication occurs in the epithelium of the small intestine and results in release of large numbers of particles in human faeces (10^{11} particles/gram). There are local (IgA) and humoral (IgM/IgG) serotype-specific and cross-reactive immune response.

Associated disease. Gastroenteritis: self-limiting diarrhoea and vomiting lasting 4–7 days; mild to severe dehydration can occur (depending on strain).

Epidemiology. Rotaviruses occur worldwide with different serotypes cocirculating, usually affecting children <2 years with a winter peak in temperate climates.

Treatment. Rehydration with oral rehydration fluid.

Prevention and control. Increased personal and water hygiene helps control outbreaks. Vaccines are under development.

CALICIVIRUSES, ASTROVIRUSES AND SMALL ROUND VIRUSES

Viruses
- Caliciviruses: RNA viruses slightly larger than picornaviruses (35 nm diameter).
- Norwalk virus: particles of 27 μm diameter; now also shown to contain the genomic RNA of a calicivirus.
- Astroviruses: 30 nm diameter particles of distinctive morphology by electronmicroscopy.

• Small, round viruses: featureless 20–30 nm round particles (possibly parvoviruses and enteroviruses.

Diagnosis. This is mainly by electronmicroscopy; ELISA and nucleic acid detection (PCR) methods are under development.

Pathogenesis and immune response. There is a short incubation period of 16–48 h with replication in epithelial cells of the villi of the small intestine.

Associated diseases. Gastroenteritis.

Treatment. Symptomatic.

HUMAN RETROVIRUSES

Human immunodeficiency virus (HIV)

Virus. A member of the Lentivirinae subfamily of the Retroviridae. There are two types: HIV-1 and HIV-2 (40% sequence homology); 5–10 subtypes. Genome: single-stranded RNA (two copies per particle).

Replication
1 Adsorption to virus-specific receptor on T-lymphocytes (CD4 cells); and receptor-mediated endocytosis (several receptors).
2 Synthesis of double-stranded complementary DNA (cDNA) by viral reverse transcriptase.
3 Integration of proviral DNA into the cellular chromosome leading to a state of chronic (persistent) infection.
4 Transcription of cDNA to yield positive-stranded RNA which acts either as messenger RNA for viral protein synthesis or as viral genomic RNA.
5 Intracytoplasmic morphogenesis.
6 Release of virus particles by budding.

Genetics. A high degree of genomic variability has been found amongst HIV isolates obtained from different persons, sequentially from the same person, and even within individual isolates (quasi-species).

Diagnosis. This is by virus isolation in cord or peripheral blood lymphocytes (in specialized laboratories only) and, routinely, by detection of HIV-specific antibody by passive particle agglutination tests (PPAT), ELISA, Western blotting (WB) and peptide blotting.

Pathogenesis and immune response. Transmission is mainly by infected blood, sexual intercourse or from mother to child during pregnancy. Chronic infection occurs of, mainly, T4 lymphocytes (T helper/inducer cells) but also of B lymphocytes, monocytes and microglial cells. This results in early impairment of various T4 cell functions followed by a decrease in T4 cell numbers and polyclonal B-cell activation. A humoral antibody response against most HIV-specific proteins develops. In later stages deficiency of both the humoral and cellular immune responses develops.

Associated diseases. Acquired immunodeficiency syndrome (AIDS) characterized by severe disease due to generalized infections with bacteria (mycobacteria), viruses (HSV, CHV, VZV), fungi (Candida, Aspergillus, Cryptococcus), protozoa (pneumocystis, toxoplasma) and associated with tumours (Kaposi's sarcoma, lymphomas).

Epidemiology. There has been worldwide spread of HIV since 1981. The main transmission routes are via infected blood, sexual contact or vertically via placenta. There is no transmission by casual contact.
 The main risk groups are:
- homosexuals;
- intravenous drug abusers;
- haemophiliac and blood transfusion recipients (before 1985);
- the sexually promiscuous;
- children born to HIV-infected mothers;
- heterosexual contacts of HIV-infected individuals.

Estimates of prevalence worldwide vary greatly; currently there are 4.5 million AIDS cases. The World Health Organization has projected that there will be a cumulative total of between 30 and 40 million cases of HIV infection worldwide by the year 2000, with more than 90% occurring in the developing countries. There are wide variations of seroprevalence rates between and within areas of different countries.

Treatment. Zidovudine (azidothymidine) limited response; other antiretroviral compounds (dideoxycytidine) are under investigation. Combination therapy will become the method of choice.

Prevention and control
- There is no antiviral cure.

- No effective vaccine has been developed; there are problems due to viral variability, latency and evasion to immune response.
- Testing of all blood and organ donors prevents transmission from those sources.
- Exposure prophylaxis is required (change of lifestyle, information campaigns, needle-exchange programmes, condom use). Specific precautions are taken in health care environments.

HTLV-1

This retrovirus is associated with adult T-cell leukaemia (ATL) and tropical spastic paraparesis (TSP).

Virus. A member of the Oncovirinae subfamily of the Retroviridae. RNA genome less variable than that of HIV; related virus, HTLV-2 (possible association with hairy T-cell leukaemia).

Diagnosis. Detection of HTLV-1 specific antibody is made by ELISA, Western blotting and other techniques.

Pathogenesis and immune response. Most infected individuals remain asymptomatic carriers for life. There is a 1% chance of developing ATL over a lifetime. HTLV-1 can invade the central nervous system. Antibody titres are normally low in carriers, but high in diseased patients.

Associated diseases. ATL and TSP.

Epidemiology. HTLV-1 is endemic in Japan and the Caribbean countries, but clusters of infection are found worldwide. It is transmitted by infected blood, sexual intercourse and breast milk.

Prevention and control. Exposure prophylaxis (see HIV). Testing of all blood donors is undertaken in some countries (in USA since 1989; not yet in UK).

ADENOVIRUSES

Adenoviruses are important causes of conjunctivitis, pharyngotracheitis and gastroenteritis but asymptomatic infections are frequent.

Virus. Non-enveloped DNA virus, 70–90 nm in diameter. There are 41 human adenovirus serotypes.

Diagnosis. This is made by: virus isolation from throat swabs or faeces in cell culture; direct detection of viral antigen in respiratory secretions by immunofluorescence; detection of viral antibody by serology (e.g. complement fixation test); detection by EM (faeces).

Pathogenesis. The incubation period is 2–5 days. Infection of the epithelial cells of the respiratory and gastrointestinal tract occurs.

Associated diseases. Pharyngitis; pharyngoconjunctival fever; acute bronchitis; pneumonia and hepatitis, especially in transplant patients; conjunctivitis; diarrhoea (types 40, 41).

Treatment. Symptomatic.

Epidemiology. Adenovirus is endemic worldwide. Infection is spread via faecal–oral route or via droplets. Epidemics have been observed among military recruits. Frequent infection occurs in transplant patients.

Prevention. Live attenuated vaccine for immunization of military personnel.

HEPATITIS VIRUSES: HEPATITIS B VIRUS (HBV)

The hepatitis B virus is a major cause of parenterally transmitted hepatitis ('serum hepatitis').

Virus. DNA virus, member of the Hepadnaviridae family, 42 nm in diameter, consisting of:
- core, DNA circular genome;
- nucleocapsid (hepatitis B core antigen, HBcAg);
- envelope (hepatitis B surface antigen, HBsAg).

Also exists as 22 nm spherical or filamentous particle consisting of HBsAg. 'e' antigen (HBeAg) is a cleavage product of the core antigen found on infected cells or free in serum.

Pathogenesis
- Parenteral infection (via infected blood or semen) with incubation period of 50–180 days.
- Viral replication in liver results in lysis of hepatocytes by cytotoxic T-cells.
- Hepatic damage reversed 8–12 weeks in ⩾90% cases; a small number of

patients develop chronic persistent or chronic aggressive hepatitis with persistence of HBsAg and HBeAg in blood (carriers).

Associated diseases. Acute and chronic hepatitis; hepatocellular carcinoma.

Diagnosis. Serological tests are made by immunoassays detecting HBsAg, HBeAg, and antibodies to HBcAg (IgM and IgG), anti-HBeAg, anti-HBsAg.

Treatment. Corticosteroids and azathioprine are used in severe cases of chronic, but not acute hepatitis. Interferon is given in cases of chronic active hepatitis.

Epidemiology
- Hepatitis B virus has worldwide distribution, with more than 200 million carriers (prevalence in north and mid-Europe and North America 0.1–0.5%; south Europe 2–5%; Africa and South-East Asia 6–20%).
- Transmission is horizontal (via blood and blood products, semen) and vertical (95% of newborns to carrier mothers become carriers if untreated; this is the main route of transmission in Asia and Africa).
- Health care personnel are at risk of infection.
- Chronic HBsAg carriers have an increased risk (approximately 200 times greater) of developing hepatocellular carcinoma.

Prevention. Effective and safe HBV vaccines are available. Initially (1982), HBsAg particles purified from serum of healthy HBsAg carriers were used as a vaccine; more recently, HBsAg is being produced from recombinant DNA in yeast cells. Neonatal vaccination prevents perinatal transmission from carrier mothers. HBeAg positive health care workers are prevented from carrying out exposure-prone procedures.

OTHER VIRUSES CAUSING HEPATITIS

Hepatitis A virus (HAV)

Virus. A picornavirus (enterovirus 72; now classified as a new genus); only one serotype.

Pathogenesis. Faeco–oral transmission occurs with an incubation period of 15–45 days (mean 30). The virus is present in blood from 2 weeks before to 1 week after jaundice; in faeces slightly longer. In most cases there is complete recovery, with specific antibody response persisting lifelong. There is no chronic disease or carrier state.

Associated disease. Acute hepatitis.

Diagnosis. This is made by serology (ELISA) testing for specific HAV IgM (acute infection) or IgG (immune status).

Treatment. Symptomatic.

Epidemiology. Hepatitis A virus is transmitted via the faecal–oral infection route and occurs mainly in children and young adults. Outbreaks occur in institutions (schools, army) linked to point sources (sewage, food). Antibody prevalence in young adults is 30–60%, higher in lower socioeconomic groups.

Prevention. A killed HAV vaccine is available.

Hepatitis C virus (HCV)

HCV is the main causative agent of parenterally transmitted non-A, non-B hepatitis.

Virus. RNA virus, a new genus of the flavivirus family, most closely related to pestiviruses. Highly variable (at least six different genotypes).

Pathogenesis and associated diseases. HCV has a similar pathogenesis to HBV; the acute hepatitis is followed by chronic hepatitis at a high incidence (30–50%). Blood can be infectious for many years.

Diagnosis. This is made by serology (ELISA) to demonstrate antibody. Polymerase chain reaction (PCR) tests are used to detect viral RNA.

Epidemiology. HCV occurs worldwide, with similar risk groups to HBV.

Prevention. Since 1991, routine testing of all blood and organ donors for HCV antibody has been undertaken to prevent transmission by transfusion or transplantation.
    Note: recently, hepatitis G virus (HGV) has been described in haemophilics. It is most closely related to HCV, but possibly a new genus.

Delta agent ('Hepatitis D virus', HDV)

Virus. Defective RNA virus, which only replicates in HBV-infected cells.

Pathogenesis and associated disease. Delta hepatitis is a coinfection or superinfection of chronic HBV infection leading to aggravation of HBV disease.

Diagnosis. Serology (ELISA) can be used to detect HDV antibody and HDV antigen.

Epidemiology. HDV infection is most prevalent in the Mediterranean area, Africa, South America and the Middle East. It has similar transmission and at-risk groups as HBV.

Hepatitis E virus (HEV)

Virus. RNA virus, most closely related to caliciviruses.

Pathogenesis and associated disease. Enterically transmitted. Disease similar to HAV disease (acute hepatitis). Specific IgM and IgG response. Severe disease and high mortality (up to 20%) in pregnant women.

Diagnosis. Serology (HEV-specific IgM and IgG, by ELISA).

Treatment. Symptomatic.

Epidemiology. Water- and foodborne epidemics in Asia, North Africa and Mexico; person-to-person spread. Seroprevalence of 40% in young adults in India.

HUMAN PAPILLOMAVIRUSES (HPV)

This is a cause of warts in humans and is associated with development of cervical cancer.

Virus. DNA virus, 55 nm in diameter. Over 70 genotypes.

Pathogenesis and associated disease. Replication takes place in epithelial cells of the skin and mucous membranes. HPV is a cause of skin warts (e.g. plantar, genital), and is associated with other skin conditions (epidermodysplasia verruciformis) and cervical carcinoma. Multiple lesions may occur in the immunocompromised (e.g. those with HIV infection).

Therapy. Surgery, cauterization or liquid nitrogen treatment may be undertaken.

Epidemiology. HPV occurs worldwide. Transmission is by close direct contact.

Prevention. Vaccines are under development.

RABIES VIRUS

This is the cause of rabies: an acute lethal infection of the central nervous system (CNS).

Virus. The rabies virus is a member of the Rhabdoviridae family. There is only one serotype. It is a bullet-shaped RNA virus (75×180 nm) surrounded by an envelope with protrusions and spikes.

Pathogenesis. Infection occurs through the bite of a rabid animal. Multiplication takes place in muscle cells and the virus migrates along peripheral nerves (Schwann sheaths) to the CNS. Replication in the CNS (mainly basal ganglia, pons, medulla) causes nerve cell destruction and inclusion bodies (Negri bodies). Migration along peripheral nerves to the salivary glands also occurs.

Associated disease. Paraesthesia; anxiety; hydrophobia; paralysis; coma. Mortality virtually 100%.

Diagnosis. This is determined by: direct immunofluorescence and histology on brain tissue of rabid animals; virus isolation in the urine; serology; reverse transcription (RT)-PCR.

Epidemiology. Rabies virus has a wide animal reservoir (e.g. foxes, skunks, racoons, bats). Bats can be healthy carriers and transmitters. There is no animal reservoir in the UK. Human infections occur mostly in Asia, Africa and South America, mainly associated with unprovoked bites from infected animals.

Therapy and control. Symptomatic treatment of Apert disease. Post-exposure, antirabies antibody plus a course of a killed vaccine should be given. Vaccination of at-risk groups (veterinary personnel, laboratory workers, travellers to endemic areas) is necessary. Spread can be controlled by quarantine of animals (UK) and vaccination of domestic animals (dogs, cats) in endemic areas.

CHAPTER 20

Basic Mycology and Classification of Fungi

CHARACTERISTICS

Fungi are eukaryotic organisms; they are distinct from plants in not containing chlorophyll. Fungi are macroscopic (mushrooms) or microscopic (moulds and yeasts). Only a few species cause human disease. Microscopic fungi are non-motile; they may grow as single cells (yeasts) or filamentous structures (mycelia), some of which may be branched.

CLASSIFICATION

Fungi are classified according to their method of sexual reproduction. Those that do not reproduce sexually are called 'fungi imperfecti'; those that reproduce sexually may be self-fertile or require strains of an opposite type to allow sexual fusion to occur.

Four groups of fungi cause human disease.

- *Yeasts*: have round or oval cells that multiply by budding, e.g. *Cryptococcus neoformans*.
- *Yeast-like fungi*: grow predominantly as yeasts that can bud. They may also form chains of elongated filamentous cells called pseudohyphae, e.g. *Candida albicans* (Plate 26).
- *Filamentous fungi*: grow as filaments (hyphae) and produce an intertwined network called a mycelium. They produce asexual spores (conidia), which may be single or multi-celled (Plate 27). Conidia are produced in long chains on an aerial hyphae (condiophore): e.g. *Aspergillus*, *Trichophyton* and *Zygomycetes* (including *Mucor*).
- *Dimorphic fungi*: have two forms of growth: filamentous at 22°C (saprophytic phase) and yeast-like at 37°C (parasitic phase); examples include *Blastomyces*, *Coccidioides*, *Histoplasma capsulatum*.

FUNGAL INFECTIONS

Fungal infections can be divided into two groups, superficial and deep mycoses (Tables 20.1 and 20.2). Most deep mycoses are opportunist infections occurring in immunocompromised patients.

Fungal infections

SUPERFICIAL MYCOSES

Fungi	Type of fungus	Principal infections	Epidemiology
Candida albicans	Yeast-like	Oral thrush, vaginitis, cutaneous candidiasis	Worldwide
Dermatophytes: Epidermophyton Microsporum Trichophyton	Filamentous	Tinea (ringworm) of skin and hair	Worldwide
Malassezia furfur	Dimorphic	Pityriasis versicolor	Worldwide, most common in the tropics

Table 20.1 Superficial mycoses.

Laboratory diagnosis
• *Direct microscopy* can be made of clinical material, including skin scrapings and sputum; the characteristic morphology facilitates identification.
• *Culture*: fungi can be isolated on most routine media (Plates 28 and 29), but may require a prolonged incubation. Antibiotic-containing selective media (e.g. Sabouraud's glucose agar), which inhibit bacterial growth, are often used. Incubation at both 37°C and 28°C facilitates the isolation of common filamentous fungi and yeasts.
• *Serology*: serological tests are available for the diagnosis of some fungal infections (e.g. candida and aspergillus), but these lack specificity and sensitivity.

Cryptococcus neoformans

• *C. neoformans* is a capsulate yeast found worldwide. Its natural habitat is soil, particularly soil contaminated with bird droppings.
• *C. neoformans* has a polysaccharide capsule which can be detected by mixing the organism in Indian ink. It can be observed in cerebrospinal fluid (CSF) by this method; other methods of detection include latex agglutination of CSF or serum.
• Infection is acquired by inhalation of the fungus. Human infections are rare.
• Primary infection occurs in the lungs and is usually asymptomatic. Acute

DEEP MYCOSES

Fungi	Type of fungus	Principal infections	Epidemiology
Aspergillus fumigatus	Filamentous	Lung infection, disseminated aspergillosis	Worldwide
Candida albicans	Yeast-like	Lung infection, oesophagitis, endocarditis, candidaemia with disseminated candidiasis	Worldwide
Cryptococcus neoformans	True yeast	Meningitis	Worldwide
Histoplasma capsulatum	Dimorphic	Lung infection	USA
Coccidioides immitis	Dimorphic	Lung infection	Central/ South America
Blastomyces dermatidis	Dimorphic	Lung infection	Africa, America
Paracoccidioides brasiliensis	Dimorphic	Lung infection	South America
Rhizopus arrhizus and related fungi	Filamentous	Rhinocerebral infection, lung infection, disseminated infection (mucormycosis)	Worldwide

Table 20.2 Deep mycoses.

pneumonia may occur with fungaemia and infection in various organs particularly the brain and meninges. Treatment is with amphotericin.

Candida

- The genus *Candida* contains a number of species, including *C. albicans* (the most frequently isolated pathogen), *C. parapsilosis* and *C. tropicalis*. *C. albicans* is a commensal of the mouth and gastrointestinal tract.
- Superficial candida infections are common and include vaginal and oral candidiasis (thrush), skin and nail infections. These infections may arise as a complication of antibiotic therapy which reduces temporarily the bacterial flora.

- Invasive candida infections may involve the gastrointestinal tract (e.g. oesophagus), lungs and urinary tract. Candidaemia may result in abscesses in various organs (e.g. brain, liver). These infections occur primarily in immunocompromised patients.
- Candida can also colonize prosthetic materials, e.g. intravascular catheters and peritoneal dialysis cannulae, resulting in septicaemia and peritonitis, respectively. Candida is a rare cause of endocarditis.

Laboratory diagnosis. By direct microscopy of appropriate clinical material for oval Gram-positive cells, some of which may be budding or producing pseudomycelia; culture (Plate 26); and serology for candida antibodies in patients with deep-seated infections.

Treatment. Topical with nystatin; parenteral therapy is with fluconazole or amphotericin B.

Malassezia furfur

- *M. furfur* is part of the normal human flora. It produces thick-walled, budding cells and curved hyphae.
- It causes pityriasis versicolor, a superficial scaly skin infection with depigmentation.

Diagnosis. By microscopy of skin scales showing yeast cells.

Treatment. Topical or oral with azole antifungals.

Aspergillus

- The genus *Aspergillus* contains a number of species, including *A. fumigatus* (the most frequent human pathogen) and *A. niger* (Plate 29).
- It is a common saprophyte worldwide, frequently found in soil and dust. Outbreaks of aspergillosis in immunocompromised patients have occurred because of construction work adjacent to hospitals which may result in the release of large numbers of spores.
- Infections are acquired by inhalation of spores (conidia), resulting in diffuse lung infection, or, occasionally, a large mycelial mass (aspergilloma).
- Infection can also spread to other sites, including the adjacent blood vessels and sinuses, or become disseminated to the liver, kidneys and brain (Plate 30). The fungus also causes chronic infections of the ear.
- *Aspergillus* infections are most frequent in immunosuppressed patients,

e.g. patients with leukaemia, transplant patients and patients with the acquired immunodeficiency syndrome (AIDS).
- *Aspergillus* is also associated with allergic alveolitis which occurs in atopic patients with recurrent exposure to aspergillus spores; symptoms include fever, cough and bronchospasm.

Laboratory diagnosis. By: direct examination of appropriate samples for branching hyphae; culture; serology (of limited value).

Treatment. Amphotericin B or itraconazole.

Dermatophytes

The dermatophytes are a group of related filamentous fungi, also referred to as the ringworm fungi which infect skin and related structures. Three clinically important genera have been described.
1 *Epidermophyton*: *E. floccosum* is the only important species (Plate 31). It infects the skin (tinea corporis), nails (tinea unguium), groin (tinea cruris) and feet (tinea pedis).
2 *Microsporum*: these infect hair and skin; *M. audouini* causes epidemic ringworm of the scalp (tinea capitis) in children. *M. canis*, which is a parasite of cats and dogs, occasionally causes ringworm in children.
3 *Trichophyton*: these infect skin, hair and nails. *T. mentagrophytes* is the most common cause of tinea pedis (Plates 32 and 33). *T. rubrum* causes severe recurrent skin and nail infections.

Epidemiology. The natural habitat is man, animals or soil; human infection results from spread from any of these reservoirs. Dermatophyte infections are found worldwide with different species predominating in various climates.

Laboratory diagnosis
- Skin scrapings, hair or nail clippings from active lesions are examined microscopically in 30% potassium hydroxide on a glass slide; the presence of hyphae confirms the diagnosis. Occasionally, the dermatophyte species can be identified by typical morphology (Plate 33).
- Samples can be cultured on Sabouraud's medium at room temperature. Subsequent species identification is based on growth rate, colony appearance and microscopic morphology (Plate 31).
- Infected hair may fluoresce under ultraviolet light (Wood's light) and is characteristic of certain infections, e.g. *M. canis*.

Treatment. Depends on the site and severity of infection. Options include topical imidazoles (e.g. clotrimazole, miconazole); oral griseofulvin, itraconazole or terbinafine.

Mucormycosis

The term 'mucormycosis' (or 'zygomycosis') refers to infections caused by a variety of filamentous fungi belonging to the order Mucorales. Medically important species include *Rhizopus arrhizus* and *Absidia corymbifera*.

Epidemiology. The Mucorales are found worldwide, in soil and decaying organic matter. Infections occur primarily in the immunocompromised. As with aspergillus, hospital outbreaks of infection have occurred in association with building work.

Infection
- Pulmonary: an often fatal infection of immunocompromised patients may result in disseminated infection (e.g. brain, liver, gastrointestinal tract).
- Rhinocerebral: an infection of the nasal sinuses may spread rapidly to involve the face, orbit, and brain. It occurs particularly in uncontrolled diabetes mellitus and is often fatal if treatment is delayed.

Laboratory diagnosis. Microscopy and culture of appropriate specimens are undertaken.

Treatment. Amphotericin, plus debridement of necrotic tissue in rhinocerebral infection.

Coccidioides immitis

C. immitis is a dimorphic fungus found in soil in hot arid areas of south-west USA, Central and South America.

Infection. This follows inhalation of arthrospores and is often subclinical. A mild, self-limiting pneumonia, often accompanied by a maculopapular rash, occurs in some patients. Severe progressive disseminated disease (e.g. meningitis, osteomyelitis) may occur, principally in immunosuppressed patients.

Laboratory diagnosis. Direct microscopy and culture of appropriate specimens (culture should only be attempted in specialized centres because of the risk of infection); serology (e.g. latex agglutination) for IgM antibodies.

Treatment. Amphotericin.

Blastomyces dermatitidis

B. dermatitidis is a dimorphic fungus found in soil in North and South America, and Africa. Man is probably infected by inhalation.

Infections. Primary pulmonary infection may be complicated by haematogenous spread to involve the skin (granulomatous ulcers), bone and joints, brain and other organs.

Laboratory diagnosis. Microscopy and culture of appropriate specimens are undertaken.

Treatment. Amphotericin B or itraconazole.

Histoplasma capsulatum

H. capsulatum is a dimorphic fungus found worldwide, but infections occur most commonly in North, Central and South America. The natural habitat is soil, particularly in sites enriched with bat droppings (e.g. caves).

Infections. Infection is acquired by inhalation of microconidia, which germinate in the lung to produce budding yeast cells. Pulmonary infection is normally self-limiting, but chronic pulmonary disease with cavitations (similar to tuberculosis) may occur in patients with underlying lung disease. Disseminated histoplasmosis (liver, bone, brain, skin) may occur, particularly in immunocompromised patients.

Laboratory diagnosis. Microscopy of stained blood films or histological sections of tissue (oral yeast-forms seen within mononuclear phagocytes); culture; or serology (e.g. complement fixation test).

Treatment. Amphotericin or itraconazole.

Paracoccidioides brasiliensis

A dimorphic fungus found in soil in Central and South America, *P. brasilien-*

sis causes pulmonary infection and mucocutaneous lesions including ulceration of the mucous membranes of the nasal and oral pharynx, which may progress to destruction of the palate and nasal septum. Disseminated infections occur with haematogenous spread to various sites, including the spleen, liver, bone and brain.

Laboratory diagnosis. Direct microscopy of pus, sputum or tissue biopsy; culture; serology (e.g. complement fixation test).

Treatment. Itraconazole, or amphotericin plus sulphadiazine.

CHAPTER 21

Parasitology: Protozoa

CLASSIFICATION

Parasites are classified into two subkingdoms (Table 21.1):
1 protozoa: unicellular organisms;
2 metazoa: multicellular organisms with organ systems.

PROTOZOA

Classification. Protozoa are classified into four groups according to structure and life cycle (Tables 21.1 and 21.2).

Amoebae

Classification. Many amoebae are human commensals (e.g. *Entamoeba coli, Endolimax nana*). *Entamoeba histolytica* is an important human pathogen. Some free-living amoebae, e.g. *Naegleria fowleri* and *Acanthamoeba* species, are opportunistic human pathogens.

Structure/physiology. Amoebae are unicellular microorganisms, with a simple two-stage life cycle:
- trophozoite: actively motile pleomorphic feeding stage;
- cyst: infective stage.

Division occurs by binary fission of the trophozoite, or by production of multiple trophozoites in a multinucleated cyst. Amoebae are motile via formation of pseudopods. Cyst formation occurs under adverse conditions.

Entamoeba histolytica

Epidemiology. E. histolytica has worldwide distribution, primarily in subtropical and tropical regions. Infected cases (symptomatic or asymptomatic carriers) act as a reservoir. Spread is usually via water or food contaminated with cysts; occasionally via oral–anal sex.

MEDICALLY IMPORTANT PARASITES

Subkingdom	Associated organisms
Protozoa	Amoebae
	Ciliates
	Coccidia (sporozoa)
	Flagellates
Metazoa	Nematodes (roundworms)
	Cestodes (tapeworms)
	Trematodes (flukes)

Table 21.1 Classification of medically important parasites.

MEDICALLY IMPORTANT PROTOZOA

Organism	Principal infections
Amoebae	
Entamoeba histolytica	Diarrhoea; dysentery; invades intenstine, with ulceration; may spread to other organs, including liver and lungs
Naegleria and Acanthamoeba	Meningitis
Ciliates	
Balatidium coli	Diarrhoea
Coccidia (sporozoa)	
Cryptosporidium parvum	Diarrhoea
Isospora belli	Diarrhoea
Toxoplasma gondii	Glandular fever syndrome
	Congenital infections, with central nervous system defects; encephalitis in the immunocompromised
Pneumocystis carinii	Pneumonia in the immunocompromised
Plasmodium species	Malaria
Flagellates	
Giardia lamblia	Diarrhoea; malabsorption
Trichomonas vaginalis	Urogenital infections
Trypanosoma species	Sleeping sickness; Chagas' disease
Leishmania species	Visceral leishmaniasis
	Cutaneous leishmaniasis

Table 21.2 Medically important protozoa.

Pathogenesis. Cysts are ingested and gastric acid promotes release of the trophozoite in the small intestine. Trophozoites multiply and may cause necrosis and ulceration in the large intestine. Invasion through the gut wall into the peritoneal cavity and blood stream spread to other organs (primarily the liver) may occur.

Clinical manifestations. Intestinal amoebiasis, with colitis and diarrhoea occurs. Liver abscesses and, less commonly, lung and brain abscesses are found.

Laboratory diagnosis. Microscopy of freshly passed stools for the presence of trophozoites and cysts, or for trophozoites in abscesses; multiple stool specimens should be examined. It is important to distinguish cysts of *E. histolytica* from cysts of non-pathogenic amoebae. Serological tests (e.g. enzyme-linked immunosorbent assay (ELISA)) can be performed but they are of limited value in endemic areas.

Treatment. Intestinal and hepatic amoebiasis are treated with metronidazole. Asymptomatic carriage may be treated with diloxanide.

Free-living amoebae

Infection may follow swimming in water contaminated with *Naegleria* and *Acanthamoeba*.

Naegleria fowleri

This is a rare cause of meningo-encephalitis. Infection follows swimming in fresh water contaminated with *N. fowleri*. The amoebae colonize the nasopharynx and then invade the central nervous system (CNS) via the cribriform plate. Cerebrospinal fluid (CSF) examination shows polymorphs and motile amoebae. Infection is frequently fatal, although some cases have been treated successfully with amphotericin.

Acanthamoeba species

This is a very rare cause of meningo-encephalitis in immunocompromised patients. Eye infections, particularly keratitis and corneal ulceration may result from contact lens cleaning solutions contaminated with *Acanthamoeba* species.

Ciliates

The only species pathogenic to man is *Balantidium coli*, a common pathogen of pigs.

Pathogenesis. *B. coli* produces proteolytic cytotoxins that facilitate tissue invasion and intestinal mucosal ulceration.

Epidemiology. *B. coli* is found worldwide, with pigs and cattle as important reservoirs. Infection is via the faecal–oral route, with occasional outbreaks following contamination of water supplies. Person-to-person spread may occur. *B. coli* cysts are ingested and trophozoites are formed, which invade the mucosa of the large intestine and terminal ileum.

Clinical manifestations. An asymptomatic carrier state can occur. *B. coli* causes gastrointestinal infection; symptoms include abdominal pain and watery diarrhoea containing blood and pus. Mucosal ulceration is rarely followed by invasive infection.

Laboratory diagnosis. Microscopy for trophozoites and cysts in faeces.

Treatment. Tetracycline or metronidazole.

Coccidia

Classification. The coccidiae include *Cryptosporidium parvum*, *Isospora belli*, *Plasmodium* species, *Toxoplasma gondii* and *Pneumocystis carinii*. They undergo asexual (schizogony) and sexual (gametogony) reproduction and have a variety of hosts, including humans.

CRYPTOSPORIDIUM PARVUM (Chapter 34)

Structure and life cycle. Mature oocytes containing sporozoites are ingested. The sporozoites are released and attach to the intestinal epithelium and mature (schizogeny). Sexual forms develop (gametogeny) and a fertilized oocyst is produced which is passed in the faeces.

Laboratory diagnosis. Microscopy of faecal specimens by a modified acid-fast stain for the presence of oocysts.

ISOSPORA BELLI

I. belli is a coccidian parasite which causes diarrhoea and malabsorption in immunocompromised patients.

TOXOPLASMA GONDII

Classification. A coccidian parasite.

Structure and life cycle (Fig. 21.1). Cats are the definitive host. Asexual and sexual cycles result in oocyst formation. Oocysts develop into trophozoites which disseminate via the blood stream, particularly to muscle and brain. Multiplication at these sites leads to tissue damage.

Epidemiology. Human infection with *T. gondii* is common, occurring worldwide. A wide variety of animals carry the organism. Humans become infected from:
- ingestion of undercooked meat contaminated with trophozoites;
- ingestion of infected oocytes from cat faeces;
- transplacental transmission;
- cardiac transplantation; the recipient receives a heart containing toxoplasma cysts.

Clinical manifestations
- Most infections are asymptomatic.
- In immunocompetent hosts: glandular fever syndrome.
- In the immunocompromised: myocarditis, choroidoretinitis meningoencephalitis.
- Congenital infection: choroidoretinitis, hydrocephalus, and intracerebral calcification.

Laboratory diagnosis
- Serology (e.g. ELISA): by rising IgG antibodies in paired sera, or IgM antibodies to differentiate between active and previous infection.
- Histology: by examination of appropriate biopsies for cysts.

Treatment. Pyrimethamine, plus sulphonamide.

PNEUMOCYSTIS CARINII

Classification. Historically, *P. carinii* has been placed with the sporozoites, but recent DNA studies suggest that it may be a fungus.

Epidemiology. Transmission is probably by droplet inhalation, but under-

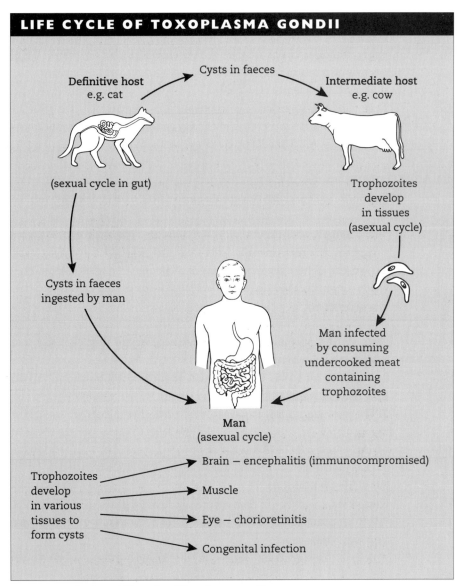

Fig. 21.1 Life cycle of *Toxoplasma gondii*.

standing of the epidemiology is incomplete. One theory suggests *P. carinii* colonizes the lungs of many individuals and infection in the immunocompromised represents reactivation.

Clinical conditions. Opportunist lung infection in the immunocompromised and in severely malnourished children is found; extrapulmonary infections occur rarely (e.g. heart, liver, kidneys and eyes).

Diagnosis. By direct examination of material, including bronchial biopsies, washings or aspirates, and open lung biopsies. Parasites are identified by histopathological stains or specific fluorescein-labelled antibodies. Serology may also be used.

Treatment and prevention. Co-trimoxazole or pentamidine are given for treatment and for prevention in certain groups of immunocompromised patients (e.g. AIDS patients, post-transplant patients).

PLASMODIUM

Plasmodia are coccidian or sporozoan parasites of erythrocytes with two hosts:
- mosquitoes, where sexual reproductive stages (gametogony) take place;
- humans or other animals, where asexual reproductive stages occur (schizogony). There are four species of plasmodia (classified in Table 21.3), which vary in detailed life cycle and clinical features, and can be separated morphologically.

Life cycle. All plasmodia share a common life cycle (Fig. 21.2) but with some important variations.

1 *Anopheles* mosquito bites a human and infectious plasmodia (sporozoites) are introduced into the blood stream.

2 The sporozoites are carried to liver parenchymal cells where asexual reproduction (schizogony) occurs in the liver to form merozoites (exoerythrocytic cycle, 7–28 days).

3 Liver hepatocytes rupture, liberating merozoites which attach to and penetrate erythrocytes.

Note: *P. vivax* and *P. ovale* have a dormant hepatic phase with sporozoites called hypnozoites, which do not divide and can result in relapse years after the initial disease.

4 Asexual reproduction occurs in the erythrocytes; merozoites are

PLASMODIA INFECTING HUMANS

Species	Fever cycle	Clinical condition
Plasmodium falciparum	36–48 h	Malignant tertian malaria
Plasmodium malariae	72 h	Quartan malaria
Plasmodium ovale	36–48 h	Benign tertian malaria
Plasmodium vivax	36–48 h	Benign tertian malaria

Table 21.3 Species of malaria infecting humans.

Protozoa 129

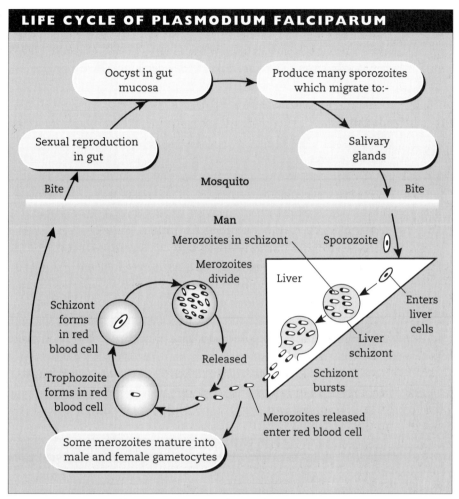

Fig. 21.2 Life cycle of Plasmodium falciparum.

released and infect further erythrocytes. Some merozoites develop within erythrocytes into male and female gametocytes. *P. falciparum* may be associated with high levels of parasitaemia (up to 30% of circulating erythrocytes); lower levels (<5%) are found with other species.

5 The mosquito bites a host and ingests mature male and female gametocytes.

6 The sexual reproductive cycle occurs in the mosquito's digestive system. Sporozoites form, migrate to salivary glands and are inoculated into a new host.

Epidemiology

- In tropical and subtropical areas plasmodia are dependent on correct

conditions for breeding of *Anopheles* mosquitos. *P. falciparum* is responsible for >80% of cases in tropical areas.
• In endemic areas, repeated infections/exposure result in relative immunity and less severe disease. Visitors to endemic areas are more severely affected.
• Transmission via contaminated blood transfusions or needle-sharing can occur rarely.

Clinical features
• The incubation period is variable (10–40 days, but may be prolonged).
• Fever/sweats: symptoms are related to release of toxins when schizonts burst and therefore intervals between bouts of pyrexia are dependent on the erythrocyte cycle of the *Plasmodium* species (Table 21.3).
• Anaemia: this is due to erythrocyte haemolysis. It is most severe with *P. falciparum* malaria and may result in haemoglobinuria ('blackwater fever').
• Cerebral malaria: high levels of parasitaemia associated with *P. falciparum* may result in erythrocyte debris blocking capillaries in the brain; the resultant hypoxia causes confusion and eventually coma, with a high mortality.

Laboratory diagnosis. Thick and thin blood films are taken. The typical morphology of the parasite within erythrocytes allows the differentiation of *Plasmodium* species.

TYPES OF MALARIA

Characteristic	P. vivax	P. ovale	P. malariae	P. falciparum
Incubation period (days)	10–17	10–17	18–40	8–11
	Sometimes prolonged for months to years			
Duration of untreated infection	5–7 years	12 months	20+ years	6–17 months
Anaemia	++	+	++	++++
Central nervous system involvement	+	±	+	++++
Renal involvement	±	+	++++	+
± to ++++, less likely to very common.				

Table 21.4 Clinical comparison of the types of malaria.

Treatment
- Chloroquine is the treatment of choice for malaria due to *P. vivax*, *P. ovale* and *P. malariae*. Supplementing treatment with primaquine is important to destroy the liver hypnozoite stages of *P. vivax* and *P. ovale*.
- Chloroquine is also the drug of choice for *P. falciparum* malaria, but chloroquine-resistance is now common in some areas of the world (e.g. South America, Central Africa, South-East Asia). Alternative drugs include quinine and the combination of pyrimethamine and sulphadoxine.

Prevention. This is important for travellers to endemic areas. Avoidance of mosquito bites is necessary (e.g. covering exposed limbs; use of mosquito nets and repellants). Prophylactic antimalarials may be taken. The exact regimen is dependent on whether resistance is present in the area being visited; examples include chloroquine, Fansidar, pyrimethamine plus dapsone (Maloprim) and chloroquine plus proguanil.

Flagellates

Classification. Human pathogenic flagellates include *Giardia lamblia* (gastrointestinal tract), *Trichomonas vaginalis* (genital tract), *Trypanosoma* species (blood/tissues) and *Leishmania* species (blood/tissues).

GIARDIA LAMBLIA (Chapter 34)

Pathogenesis and life cycle. Cysts are ingested and gastric acid stimulates the release of trophozoites in the small intestine, which then multiply by binary fission. Trophozoites attach to the intestinal villi by a sucking disc (Plate 34). Inflammation of the epithelium may occur, but systemic invasion is rare. Throphozoites divide by binary fission. Cyst formation occurs as the organisms move through colon.

Epidemiology. *Giardia* occurs worldwide. Transmission is via ingestion of contaminated water or food, or direct person-to-person spread via the faecal–oral route.

Clinical manifestations. Asymptomatic carriage is common. Active infection causes diarrhoea and malabsorption occasionally.

Laboratory diagnosis. Stool microscopy for cysts and trophozoites (Plate 34). If clinical suspicion is high and stool examination is negative, duodenal aspiration or biopsy may be helpful.

TRICHOMONAS VAGINALIS

Structure. *T. vaginalis* is a pear-shaped protozoa with four flagella. No cyst stage has been recognized.

Epidemiology. It has a worldwide distribution, with sexual intercourse the primary method of spread.

Pathogenesis/clinical conditions. *T. vaginalis* is a parasite of the human urogenital system; there are no animal reservoirs. Infection results in a watery vaginal discharge, although extensive inflammation and erosion of the epithelium can occur. Males are usually asymptomatic carriers, however, occasionally, urethritis and prostatitis can occur.

Laboratory diagnosis. This is by: microscopy of vaginal or urethral discharge for trophozoites; isolation in special media.

Treatment. Metronidazole.

TRYPANOSOMA SPECIES

Trypanosomes are haemoflagellates and live in the blood and tissue of human hosts. The life cycle involves two hosts: blood-sucking insects and mammals. It causes two diseases.
- African trypanosomiasis (sleeping sickness) caused by *T. brucei gambiense* and *T. brucei rhodesiense*, and is transmitted by the tsetse fly.
- American trypanosomiasis (Chagas' disease) caused by *T. cruzi*, and is transmitted by the reduviid bug.

Trypanosoma brucei gambiense (West African sleeping sickness)

Life cycle

1 The infective stage (trypomastigote) is present in salivary glands of the tsetse fly. The trypomastigote has a flagellum and an undulating membrane along the length of the body allowing motility.

2 Trypomastigotes enter the host via the insect bite and reach the lymphatic system, blood and central nervous system (CNS). Non-flagellate forms (amastigotes) develop in some tissues (e.g. heart and muscle).

3 Tsetse flies feed on the infected host and take in trypomastigotes which multiply in the intestine; epimastigotes subsequently form in salivary glands and develop into infective trypomastigotes.

4 The tsetse fly remains infective for life.

Epidemiology. Limited to West and Central Africa. Domestic animals and asymptomatic humans are the main reservoirs.

Clinical condition. The incubation period is a few days to several weeks. A nodule or chancre can form at the site of the bite, which usually resolves spontaneously. The parasite enters the blood stream and results in lymphadenopathy, irregular fever and myalgia. CNS involvement (sleeping sickness) may occur with lethargy and encephalitis, leading to convulsions, hemiplegia, coma and occasionally death.

Diagnosis. Microscopy of thick and thin blood films, lymph node aspirates and cerebrospinal fluid for trypomastigotes; serology (e.g. ELISA).

Treatment. Suramin; melarsoprol if CNS involvement.

Trypanosoma brucei rhodesiense (East African sleeping sickness)

Life cycle. The life cycle is similar to *T. brucei gambiense*, with trypomastigote and epimastigote stages. Transmission is by the tsetse fly.

Epidemiology. It is found in East Africa. Domestic and game animals are the main reservoirs. Asymptomatic human carriers are not normally a source of infection.

Clinical conditions. It has a shorter incubation than *T. brucei gambiense*. The illness is more severe and progresses rapidly and may involve the heart, kidney and CNS. Mortality is greater than for the West African form.

Diagnosis and treatment. As above.

Trypanosoma cruzi (Chagas' disease)

Life cycle
- The life cycle is similar to other trypanosomes, except when the infected bug bites humans, trypomastigotes are released simultaneously in the faeces; these enter the wound following scratching.
- Spread is via the lymphatic and blood systems and results in invasion of many organs, including liver, heart, muscle and brain.
- Within host cells, trypomastigotes transform into amastigotes, which multiply by binary fission and form either further amastigotes or trypomastigotes; the latter are ingested by a feeding insect, multiply in the intestine and are then passed in the faeces.

Epidemiology. *T. cruzi* is found in North, Central and South America. Reservoirs are man and domestic animals.

Clinical disease. Chagas' disease may be asymptomatic, acute or chronic. A painful nodule may form at the bite. Acute infection results in high fever, erythematous rash, oedema and myocarditis. Chronic disease may develop years after the initial infection. Organisms proliferate in various organs, including the brain, spleen, liver, heart and lymph nodes, resulting in lymphadenopathy, hepatosplenomegaly, myocarditis and cardiomegaly. Granulomas and cysts may form in the brain. Cardiac disease is the most common presentation, with congestive cardiac failure.

Laboratory diagnosis. Microscopy of biopsies of affected tissues; thick blood films for trypomastigotes in the acute phase; serological tests (e.g. ELISA).

Treatment. Nifurtimox.

Prevention. Insecticides; avoidance of insect bites; improvement in living conditions.

LEISHMANIA SPECIES

Leishmania species are obligate, intracellular parasites transmitted to mammalian hosts by sandflies. An estimated 12 million people are infected worldwide.

Classification. There are three main species that produce human disease: *Leishmania donovani*, *L. tropica* and *L. braziliensis*. These species are associated with various clinical syndromes (Table 21.5).

Structure. Leishmaniae are flagellated protozoa.

Life cycle
1 A promastigote stage is present in the saliva of infected sandflies and is inoculated into the bite site.
2 Promastigotes lose their flagella, enter the amastigote stage, are engulfed by tissue macrophages, and carried via the reticulo-endothelial system to the bone marrow, spleen and liver. Amastigotes multiply resulting in tissue damage.
3 Amastigotes are taken up by the sandfly during feeding, develop into

LEISHMANIA SPECIES AND INFECTIONS

Organism	Reservoir	Clinical condition	Location	Average incubation period
L. donovani	Man or dog	Visceral leishmaniasis (kala-azar)	Africa Asia	3 months
L. tropica	Man or dog	Cutaneous leishmaniasis (oriental sore)	Africa Asia Mediterranean	1–2 months
L. major	Rodents			
L. braziliensis	Sloths and related species	Mucocutaneous leishmaniasis	Central and South America	a few weeks to months

Table 21.5 Syndromes caused by *Leishmania* species.

promastigotes, multiply in the mid-gut, and then migrate to the salivary glands.

Visceral leishmaniasis

Caused by *L. donovani*.

Clinical conditions. It can be asymptomatic. Fever, weight loss and diarrhoea are present, with hepatosplenomegaly and renal involvement. Darkening of the skin, mainly the face and hands (kala-azar), malabsorption and anaemia may occur. Most infections resolve without therapy.

Diagnosis. Histological examination of appropriate tissue biopsies; bone marrow or lymph nodes for amastigotes; serology.

Treatment. Sodium stibogluconate or pentamidine isethionate.

Cutaneous leishmaniasis

Caused by *L. tropica* and *L. major* complexes.

Clinical conditions. Cutaneous disease with a papule at the bite site, followed by necrosis of epidermis and ulceration. Infections remain localized, although lesions may be multiple.

Diagnosis. Histological examination for amastigotes in smears or biopsy material; serology.

Treatment. Sodium stibogluconate.

Mucocutaneous leishmaniasis

Caused by *L. braziliensis*.

Clinical conditions. Infection mainly involves the mucous membranes of the upper respiratory tract (palate, nose) and related tissue, with ulceration, tissue destruction and resulting disfigurement.

Diagnosis. Histological examination of biopsies; serology.

Treatment. Sodium stibogluconate or amphotericin B.

CHAPTER 22

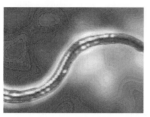

Parasitology: Metazoa (Helminths)

NEMATODES

The most common nematodes of medical importance are: *Ascaris lumbricoides*, *Enterobius vermicularis*, *Toxocara canis* and *Toxocara cati* (Table 22.1).

Ascaris lumbricoides

Structure and life cycle (Fig. 22.1)
- Adult worms are cylindrical; the female is up to 35 cm long, the male up to 30 cm long.
- Following ingestion of an infective egg, larval worms are released in the duodenum, which penetrate the intestinal wall to reach the blood stream, passing via liver and heart to the pulmonary circulation. These larval worms pass into the alveoli, migrate via the bronchi, trachea and pharynx, are swallowed and return to the small intestine.
- Male and female worms mature and mate in the intestine; up to 200 000 eggs per day are produced by the female and passed in stools.

Epidemiology. *Ascaris lumbricoides* is found throughout the world, but particularly in areas of poor sanitation where faecal–oral transmission may occur. Eggs can survive for long periods (several years), making *Ascaris* the most common pathogenic helminth worldwide.

Clinical manifestations/complications. Victims are asymptomatic when the worm load is light. Vomiting and abdominal discomfort may occur.
  Complications include:
- pneumonitis (following larval migration in the lungs);
- intestinal obstruction with mature worms (children);
- perforation of the intestine and hepatic abscesses (rare).

Laboratory diagnosis. Stool examination for presence of fertilized or unfertilized eggs. An adult worm is occasionally passed in faeces.

MEDICALLY IMPORTANT NEMATODES

Organism	Notes on disease/infection
Ascaris lumbricoides (Roundworm)	Larval worms migrate via bronchi; adult worms may cause intestinal obstruction in children
Enterobius vermicularis (Pin- or threadworm)	Perianal itching
Trichinella spiralis	Peri-orbital oedema; myalgia; fever
Toxocara canis	Larvae migrate to liver, lungs and eyes (visceral larva migrans)
Trichuris trichiura	Anaemia, intestinal irritation

Table 22.1 Medically important nematodes.

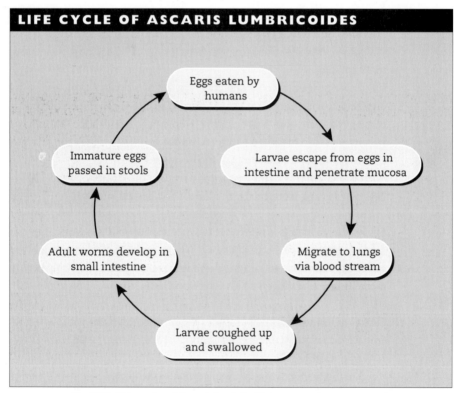

Fig. 22.1 Life cycle of *Ascaris lumbricoides*.

Treatment and prevention. Treatment is with mebendazole or piperazine. Control is by improved sanitation and avoidance of food contaminated with human faeces.

Enterobius vermicularis

Structure and life cycle (Fig. 22.2)
- The female worm (10 mm × 0.4 mm) has a pointed tail; the male worm (3 mm × 0.2 mm) has a curved end. Infection follows ingestion of an embryonated egg.
- Larvae hatch in the small intestine and migrate to the large intestine where they mature into adults in approximately 1 month.
- The male fertilizes the female who migrates to the perianal area and lays eggs; these eggs become infective within hours.
- Perianal irritation leads to scratching; eggs contaminate hands, particularly nails, resulting in autoinfection or transmission to a new host.

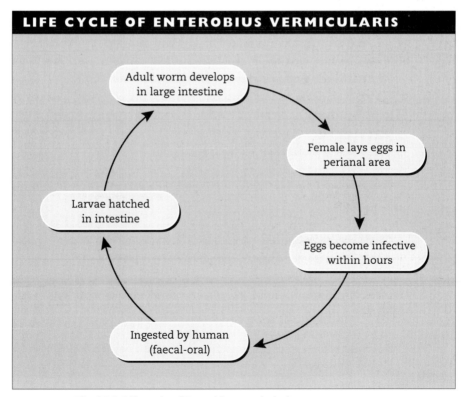

Fig. 22.2 Life cycle of *Enterobius vermicularis*.

Epidemiology. This is a common parasitic infection, with 500 million cases worldwide, mainly in temperate regions. Person-to-person spread is responsible for transmission.

Clinical manifestations and complications. Individuals are frequently asymptomatic. Patients allergic to worm secretions may experience perianal pruritus. In heavily infected females, worm migration into the vagina may occur.

Laboratory diagnosis. By microscopy; the best method is a cellotape slide applied around the perianal region to pick up eggs. Adult worms may be found in stool specimens.

Treatment and prevention. Mebendazole is given to treat the entire family simultaneously. Improved personal hygiene is advised, including cutting of fingernails and regular washing of bedclothes and towels.

Toxocara canis and T. cati

Structure and life cycle
- *T. canis* and *T. cati* are parasites of dogs and cats respectively. Male and female worms develop in the intestine and produce eggs which are ingested by other dogs or cats and the cycle continues.
- Man is an unintentional host; on ingestion, the eggs hatch into larval forms which can penetrate the intestinal mucosa and migrate to various organs, particularly the lung, liver and eyes. Larvae are unable to develop further and granuloma formation results.

Epidemiology. It is associated with infected dogs and cats; worldwide distribution.

Clinical manifestations and complications. Toxocariasis may result in pneumonitis, hepatosplenomegaly, eosinophilia and retinitis.

Laboratory diagnosis. Serology by enzyme-linked immunosorbent assay (ELISA).

Treatment and prevention. Often unnecessary; thiabendazole and steroids are used in severe cases. Young pets should be treated routinely ('deworming').

Trichuris trichiura (whipworm)

Structure and life cycle
- Ingested eggs develop into the larvae in the small intestine which migrate to the large intestine and mature into adult worms. The fertilized female worm produces eggs.
- Eggs are excreted and mature in the soil.

Epidemiology. Trichuris trichiura has a worldwide distribution, particularly in areas with poor sanitation. No animal reservoir has been identified.

Clinical manifestations and complications. Patients are usually asymptomatic, but worm infection may produce abdominal distention and diarrhoea with weight loss; anaemia, eosinophilia, bloody diarrhoea and anal prolapse are seen occasionally.

Laboratory diagnosis. Stools are examined for characteristic eggs.

Treatment. Mebendazole is the drug of choice.

Prevention. Good personal hygiene and adequate sanitation can help to prevent infection.

Ancylostoma duodenale and Necator americanus

Structure and life cycle
- Infective larvae penetrate intact skin, enter the circulation and are carried to the lungs; following maturation, larvae are expectorated, swallowed and develop into adult worms in the small intestine.
- The adult worms lay eggs which are excreted and, in hot, humid conditions, develop rapidly into infective larvae which may infect a new host by penetration of exposed skin, particularly bare feet.

Epidemiology. Found primarily in subtropical and tropical regions.

Clinical manifestations and complications
- There is a rash at the entry site; occasionally pneumonitis occurs during larval migration.
- Adult worms produce gastrointestinal-related symptoms, including vomiting and bloody diarrhoea.

- Chronic anaemia may result from blood loss, particularly in the malnourished.

Laboratory diagnosis. Stool microscopy for eggs.

Treatment and prevention. Mebendazole for treatment; preventative measures include improved sanitation and wearing of shoes in endemic areas.

Strongyloides stercoralis

Structure and life cycle
- *Strongyloides stercoralis* are similar to hookworms, except that the eggs hatch into larvae (Plate 35) in the intestinal mucosa before being passed in faeces. The larvae mature directly into infective larvae able to penetrate skin, or enter a non-parasitic cycle outside the human host.
- Larvae may also reinfect the host by penetrating the intestinal mucosa (autoinfection); this is more common in immunocompromised patients.
- Infection may be quiescent for up to 30 years and then reactivate when the immune system is compromised (e.g. organ transplantation) resulting in disseminated infection.

Epidemiology. They are found in warm, moist environments, similar to hookworm.

Clinical manifestations. Pneumonitis, diarrhoea, malabsorption and gut ulceration may occur. Immunosuppressed patients may develop severe infection affecting multiple organs.

Laboratory diagnosis. Stool microscopy for larvae; serology.

Treatment and prevention. Thiabendazole; adequate sanitation aids prevention.

Trichinella spiralis

Structure and life cycle
- Human infection occurs accidentally as this parasite is found primarily in carnivorous animals. Humans may become affected following ingestion of encysted larvae in undercooked meat, particularly pork.
- Larvae are released in the small intestine where they mature into adult worms which produce further larvae. The larvae penetrate the intestinal

wall and are released into the circulation and become encysted in various tissues, particularly striated muscle.
• The life cycle is completed in animals who are infected following ingestion of striated muscle containing encysted larvae.

Epidemiology. Worldwide distribution; associated with eating pork.

Clinical manifestations. Trichinellosis presents with muscle pain, tenderness and periorbital oedema. Myocarditis, encephalitis and pneumonitis may occur. Infection is occasionally fatal.

Laboratory diagnosis. Eosinophilia, the presence of encysted larvae in implicated meat or biopsied muscle from the patient and serology can be used.

Treatment and prevention. Mebendazole; pork should be cooked thoroughly.

Wuchereria bancrofti and Brugia malayi

These are transmitted by mosquitoes.

Structure and life cycle
• Immature larvae are introduced following a bite from an infected mosquito; they migrate via lymphatics to regional lymph nodes.
• Adult worms mature in the lymphatics, fertilization occurs and larval microfilariae are produced. These enter the circulation and are ingested by feeding mosquitoes.
• In the mosquito, larvae migrate via the stomach to the mouthpiece to complete the cycle.

Epidemiology. The worms are found in tropical and subtropical Africa, Asia and South America.

Clinical manifestations and complications. Early signs include influenza-like illness, lymphangitis, lymphadenopathy. Chronic infection results in blockage of the lymphatics with peripheral oedema, and eventually limb fibrosis (elephantiasis).

Laboratory diagnosis. By detection of microfilariae in blood films (taken at night as microfilariae show nocturnal periodicity), and serology.

Treatment and prevention. Diethylcarbamazine; mosquito control prevents spread.

Loa loa

Structure and life cycle
- *Loa loa* is similar to *Wuchereria bancrofti*. The insect vector is *Chrysops*, the mango fly.
- In humans, larvae mature and migrate subcutaneously. Fertilization results in microfilariae, which pass into the blood stream and are ingested by feeding flies.

Epidemiology. *Loa loa* infection is found in tropical Africa.

Clinical manifestations. There is swelling where the worms migrate subcutaneously. Adult worms may be seen migrating under the conjunctiva, hence the name 'eye-worm'.

Laboratory diagnosis. By detection of microfilariae in blood and serology.

Treatment and prevention. Treatment is with diethylcarbamazine and by surgical removal of localized worms. Prevention is by protection from insect bites.

Onchocerca volvulus

Infection follows a bite from black flies which breed near fast-flowing rivers. Larvae develop into adult worms in subcutaneous nodules. *Onchocerca volvulus* is found in Africa and Central Southern America. Chronic infection results in atrophic skin and may cause blindness due to microfilariae migrating to the eyes. Diagnosis is by skin biopsy or slit-lamp examination of the anterior chamber of the eye for microfilariae. Diethylcarbamazine is the drug of choice.

CESTODES

Tapeworms are ribbon-like and up to 30 feet in length. A head or scolex has suckers with a crown of hooklets, facilitating attachment. Segments of tapeworms are called proglottids. All tapeworms are hermaphrodite, with both male and female reproductive organs present in each segment. Food is absorbed from the host intestine through the body wall. Most

Cestodes

CLINICALLY IMPORTANT CESTODES

Cestode	Common name	Reservoir for larvae	Reservoir for adult worm
Taenia solium	Pork tapeworm or cysticercosis	Pigs	Humans
Taenia saginata	Beef tapeworm	Cattle	Humans
Diphyllobothrium latum	Fish tapeworm	Crustacea, fish	Humans, cats, dogs
Echinococcus granulosus	Hydatid cyst	Humans	Canines
Hymenolepis nana	Dwarf tapeworm	Rodents, humans	Rodents, humans

Table 22.2 Clinically important cestodes.

tapeworms have complex life cycles, with intermediate hosts. Clinically important cestodes are listed in Table 22.2.

Taenia solium

Structure and life cycle (Fig. 22.3). Humans ingest pork containing larvae (cysticerci) and the larvae develop in the small intestine into the adult form. The worm produces proglottids which mature sexually, producing eggs which pass in the faeces. Pigs ingest the eggs and larval forms develop which disseminate via the circulation to muscle to produce cysticerci.

Epidemiology. Infection is related to eating undercooked pork. It is common in Africa, India, South-East Asia and Central America.

Clinical manifestations. Abdominal discomfort and diarrhoea.

Diagnosis. Stool microscopy for proglottids and eggs; serology.

Treatment and prevention. Treatment is with praziquantel. Pork should be cooked thoroughly and improved sanitation is important for prevention.

Cysticercosis

Humans may become infected with the larval stage of *T. solium* following ingestion of eggs from human faeces. Larvae penetrate the intestinal wall, enter the circulation and are carried to muscle, brain, lungs and occasionally the eyes. The inflammatory response to the presence of these larvae in

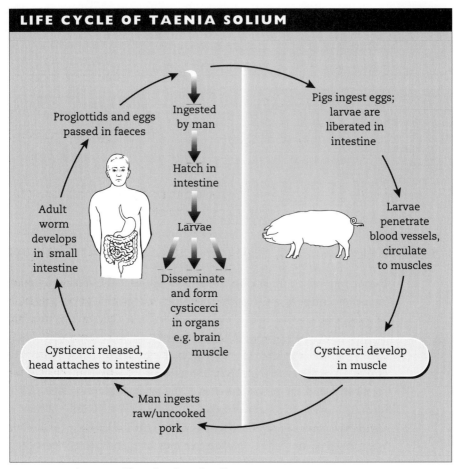

Fig. 22.3 Life cycle of *Taenia solium*.

human tissues eventually results in calcification. Infection often subclinical but may present with neurological or ophthalmic manifestations.

Laboratory diagnosis. By: X-ray; computed tomography; histology; serology (ELISA).

Treatment. Praziquantel.

Taenia saginata

Structure and life cycle. The life cycle is similar to *Taenia solium*, infection resulting from ingestion of cysticerci present in insufficiently cooked beef.

Human faeces contaminated with eggs are ingested by cattle and develop in beef.

Epidemiology. Worldwide distribution.

Clinical manifestations. Abdominal discomfort.

Laboratory diagnosis. Proglottids and eggs are detected by stool microscopy.

Treatment and prevention. Treatment with praziquantel. Improved sanitation aids prevention. Beef should be cooked thoroughly.

Diphyllobothrium latum (fish tapeworm)

Structure and life cycle. There are two intermediate hosts, freshwater crustaceae and freshwater fish. Infection occurs in humans following ingestion of infected freshwater fish. The adult worm develops in the intestine, attaching to the mucosa via lateral grooves. Proglottids develop and produce eggs. In fresh water the eggs develop into larvae, which infect crustaceae. Fish eat the crustaceae, and larvae develop which are infective for humans.

Epidemiology. Infection occurs in temperate climates, particularly Scandinavia.

Clinical manifestations. Diarrhoea and occasionally intestinal obstruction. Anaemia and vitamin B12 deficiency occur.

Treatment and prevention. Praziquantel is the drug of choice. Avoid eating raw and undercooked fish.

Echinococcus granulosus

Structure and life cycle
- The adult tapeworms are found in canines, normally dogs or foxes. In the canine intestine the adult tapeworms produce eggs, which are passed in the faeces.
- Herbivores, the normal intermediate hosts, ingest the eggs and larvae develop in various organs, resulting in cyst formation. Carnivores become infected after consuming contaminated carcasses.

- Human infection occurs when man becomes the accidental intermediate host. Eggs are ingested, larvae develop and penetrate the intestinal wall; hydatid cysts develop, principally in the liver. These cysts accumulate fluid and enlarge over several years; a germinal layer surrounds the cyst, which produces immature tapeworm heads.

Epidemiology. *Echinococcus granulosus* is associated with sheep farming in Europe, Australasia, South America and USA. Human infection follows ingestion of water or vegetation contaminated with eggs.

Clinical manifestations. Cysts in the liver grow over many years, resulting in hepatomegaly and jaundice. Cysts in the lung, bone and brain may result in various clinical presentations.

Laboratory diagnosis. Radiology; microscopy of cyst contents; serology.

Treatment and prevention. Treatment is by resection of cysts or aspiration and mebendazole. Avoid ingestion of contaminated food.

Hymenolepsis nana

This is a common, often symptomless infection occurring particularly in Asia. *Hymenolepsis nana* is a small (2–5 cm) worm. Eggs are produced in the small intestine and may reinfect the same host or spread to other hosts by the faecal–oral route. Heavy infection may result in diarrhoea, abdominal discomfort and weight loss. Treatment is with praziquantel.

TREMATODES

Except for schistosomes, trematodes (flukes) are flatworms with oral and ventral suckers. Most flukes are hermaphrodite, containing both male and female reproductive organs in a single body. Schistosomes differ in having separate male and female worms.

All flukes require a reservoir host and intermediate host for completion of their life cycle. The intermediate hosts are molluscs, in which the sexual reproduction cycle occurs. Some flukes need a further intermediate host. The medically important trematodes are summarized in Table 22.3.

Fasciola hepatica (sheep liver fluke)

Structure and life cycle

MEDICALLY IMPORTANT TREMATODES

	Common name	Intermediate host	Vector	Reservoir host
Fasciola hepatica	Sheep liver fluke	Snail	Aquatic plants (e.g. watercress)	Sheep, cattle
Opisthorchis sinensis (formerly *Clonorchis sinensis*)	Chinese liver fluke	Snail, freshwater fish	Uncooked fish	Dogs, cats, humans
Paragonimus westermani	Lung fluke	Snail, crayfish	Uncooked crabs, crayfish	Humans
Schistosoma species	Blood flukes	Snail		Humans

Table 22.3 Medically important trematodes (flukes).

- *F. hepatica* is a parasite of herbivores (sheep, cattle) and occasionally humans.
- Human infection starts with ingestion of watercress or similar vegetation contaminated with encysted metacercariae which excyst in the duodenum. The larval flukes migrate through the duodenal wall, across the peritoneal cavity and into the bile ducts via the liver. They mature into adult worms and produce eggs which are passed in faeces.
- Snails become infected and act as intermediate hosts. Cercariae (infective larvae) develop and contaminate watercress where they encyst to form metacercariae.

Epidemiology. *F. hepatica* occurs worldwide, especially in sheep-rearing areas.

Clinical manifestations. Migration of the larval worm via the liver causes hepatomegaly. Worms in the bile duct may cause biliary obstruction and eventually cirrhosis.

Laboratory diagnosis. Stool microscopy for eggs.

Treatment and prevention. Praziquantel. Avoid eating watercress grown in areas frequented by sheep and cattle.

Opisthorchis sinensis (Chinese liver fluke; formerly known as Clonorchis sinensis)

Structure and life cycle. O. sinensis has a similar life cycle to F. hepatica except there are two intermediate hosts. Eggs are eaten by a snail and develop into adult worms; free-swimming cercariae leave the snail and enter freshwater fish by penetrating under scales where they develop into metacercariae. Man becomes infected by eating contaminated raw fish. Larval forms penetrate the duodenum, migrate to the bile duct and develop into adult worms.

Epidemiology. O. sinensis is found in China, Japan, Korea and South-East Asia.

Clinical manifestations. People are usually asymptomatic, but severe infections can result in hepatomegaly, biliary obstruction and jaundice.

Laboratory diagnosis. Stool microscopy for eggs.

Treatment and prevention. Praziquantel is the drug of choice. Avoid eating uncooked fish. Improved sanitation is important in prevention.

Paragonimus westermani (lung fluke)

Structure and life cycle
- Embryonated eggs in water produce miracidia, which penetrate snails and develop into cercariae. The latter infect crabs and crayfish.
- Humans become infected by eating infected crabs or crayfish. The larvae hatch in the duodenum, migrate through the intestinal wall, across the peritoneal cavity and eventually to the lungs via the diaphragm.
- The worms reside within fibrous cysts in the lungs and produce eggs which then appear in the sputum or, when swallowed, in faeces.

Epidemiology. P. westermani is found in Asia, Africa, India and South America.

Clinical manifestations. Adult flukes in lungs may result in cough and chest pain; lung fibrosis may develop. Larval migration can also cause disease in other organs, particularly the brain.

Laboratory diagnosis. Microscopy of sputum and faeces for eggs and serology.

Treatment and prevention. Praziquantel. Avoid eating uncooked freshwater crabs and crayfish; improve sanitation.

Schistosomes

Schistosomes associated with human infection (schistosomiasis) include *Schistosoma mansoni*, *S. japonicum* and *S. haematobium* (Table 22.4). Schistosomiasis is an important infection with over 200 million people infected worldwide.

Structure and life cycle (Fig. 22.4)
- Schistosomes have separate male and female worms.
- Infective cercariae penetrate the skin, enter the circulation and develop into male and female worms in the veins of the intestine (*S. mansoni*, *S. japonicum*) or bladder (*S. haematobium*). The female worm produces eggs which penetrate the intestinal or bladder wall and are passed in faeces or urine. Ova reach fresh water, hatch into larvae and infect snails (Plate 36). An asexual cycle results, with formation of cercariae which are released into water.

Clinical manifestations
- Skin penetration may result in local dermatitis.

Schistosoma species	Location in human	Epidemiology	Laboratory diagnosis
S. mansoni	Inferior mesenteric vein	Endemic in Africa, Middle East and South America	Stool for eggs with lateral spine (Plate 36) Serology
S. japonicum	Superior and inferior mesenteric vein	China and Japan	Stool for eggs Rectal biopsy for eggs Serology
S. haematobium	Inferior pelvic system	Egypt and Middle East	Urine for eggs with terminal spine Bladder biopsy for eggs Serology

Table 22.4 Diagnosis of schistosome infection.

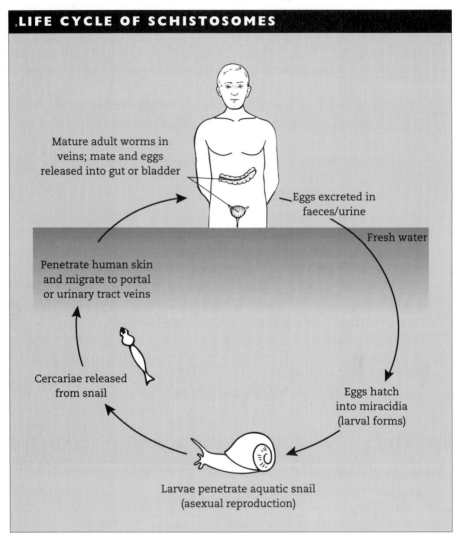

Fig. 22.4 Life cycle of schistosomes.

- Egg release is associated with haemorrhage, normally minor, into the bladder (haematuria) or intestine (blood in faeces).
- Egg entrapment in the bladder or intestinal wall may result in a chronic inflammatory reaction with subsequent fibrosis:
    (a) fibrosis and calcification of the bladder wall; obstruction of ureters, resulting in hydronephrosis and renal failure;
    (b) fibrosis of the portal vein, resulting in portal hypertension.
- Egg deposition may occur in other organs, e.g. the liver (fibrosis, portal hypertension), lungs and brain.

Laboratory diagnosis. Table 22.4 gives details of laboratory diagnosis of schistosomes.

Treatment and prevention. Praziquantel. Prevention is by improvement of sewage disposal; education (e.g. bare feet in endemic areas); and the use of molluscicides.

CHAPTER 23

Host–Parasite Relationships

ECOLOGY

Microorganisms have developed characteristics that allow them to survive and grow in a wide variety of environments and are found in virtually every ecological niche. They play a central role in global ecology, providing both an important food source and facilitating the degradation of organic matter.

- Microorganisms that live on inanimate organic material are described as *saprophytes*.
- Microorganisms that grow in or on a living host, obtaining nutrients from the host are referred to as *parasites*; they may be beneficial to the host (*symbiosis*), e.g. in the stomach of ruminants they aid in the degradation of cellulose—a vital step in the nutrition of such animals.

NORMAL FLORA

Parasitic microorganisms associated with mammals are often referred to as normal flora or commensals. They are important in preventing infection; reduction in normal flora (e.g. by antibiotic therapy) may result in overgrowth of potential pathogens.

Commensals are well adapted to their environment, e.g. the dry, acidic conditions found on the skin, and can multiply without causing host damage. However, commensal bacteria can cause infection, normally when some aspect of host defence is compromised; for example:

- *Staphylococcus epidermidis*, a skin commensal, is a cause of intravascular catheter infections (p. 297);
- Viridans streptococci (oral commensal bacteria), may cause bacterial endocarditis (p. 276).

PATHOGENIC MICROORGANISMS

Microorganisms which frequently cause infections are referred to as *pathogenic*. However, they may be present on the host without resulting in infection; for example:

- *Neisseria meningitidis*, an important cause of meningitis, is a commensal of the pharynx of about 5% of adults;
- *Streptococcus pneumoniae*, the most frequent cause of bacterial pneumonia, is a pharyngeal commensal in many individuals;
- *S. aureus*, an important cause of skin infection and deep abscesses, is a nasal commensal of about one-third of individuals.

INFECTION

- Infection describes the clinical manifestations that occur when a microorganism invades a host.
- When the manifestations are minor or imperceptible, infections are often termed *subclinical*.
- *Latent infection* results when pathogens persist in the body without evoking a clinical response; periodically, overt infections occur following a change in the patient's immune state (e.g. cytomegalovirus infection in transplant patients; herpetic cold sores).
- *Chronic* or *persistent infections* occur when pathogens are not eradicated completely and continue to evoke a clinical response (e.g. leprosy; helminth infections).

MICROBIAL STRATEGIES

Microorganisms have developed a variety of mechanisms (virulence factors) to combat host defences and promote transmission to new hosts.

Transmission

Ability to survive outside the host is an important factor in transmission of organisms. Upper respiratory tract viruses (e.g. rhinoviruses) survive poorly outside the host and successful transmission relies on the production of large quantities of infectious particles. Other organisms can survive outside the host, e.g. *Mycobacterium tuberculosis* and do not require large numbers of infectious particles for efficient transmission.

Attachment

Many organisms have specific factors that allow attachment to mucosal surfaces, e.g. the cell surface lipoteichoic acid of group A β-haemolytic streptococci and pili of *N. gonorrhoea*.

Replication and combating host defences

- The haemolysins and leucocidins of streptococci and staphylococci reduce function (chemotaxis, phagocytosis) and viability of phagocytic cells.
- Surface components of some pathogenic bacteria inhibit the activation and deposition of complement, preventing opsonization, e.g. the capsule of *S. pneumoniae*.
- Organisms have a number of strategies for neutralizing the effect of antibody. These include: production of soluble antigen, which neutralizes the antibody in the fluid phase preventing binding to the microbial surface; binding of antibody in an inverted fashion, e.g. staphylococcal F_c receptor; direct destruction of antibodies by proteases, e.g. IgA protease of *Pseudomonas aeruginosa*; variation in antigenic structure requiring the host to synthesize new specific antibody, e.g. antigenic drift of influenza viruses.
- Intracellular survival: some microorganisms can survive and multiply within phagocytic cells. They avoid or neutralize the oxygen-dependent killing mechanism of phagocytic cells or prevent lysosomal fusion (e.g. mycobacteria, brucella, legionella, listeria, and many viruses).

Toxigenicity

Organisms can produce a variety of toxins, of which there are two main types, exotoxins and endotoxins; their main properties are shown in Table 23.1.

ENDOTOXINS

Endotoxins (lipopolysaccharides (LPS)) are important in the pathogenesis

EXOTOXINS AND ENDOTOXINS	
Exotoxins	Endotoxins
• Mainly produced by Gram-positive bacteria • Heat-labile proteins • High potency • Strong antigenicity • Neutralized by antitoxin • Often possess specific mechanisms of action	• Lipopolysaccharides—part of Gram-negative cell wall • Heat-stable • Liberated when Gram-negative bacteria lyse • Non-specific effect

Table 23.1 Properties of exotoxins and endotoxins.

of Gram-negative septic shock (endotoxic shock). The LPS molecule activates the complement and cytokine pathways, releasing activated components which increase vascular permeability and result in shock; the clotting and fibrinolytic cascades are also activated, resulting in haemorrhage and thrombosis (disseminated intravascular coagulopathy).

EXOTOXINS

Some exotoxins provide obvious benefits to the organism in protecting against host defences or promoting transmission; the benefit of other toxins is harder to understand. Examples of exotoxins in disease:

- *diphtheria toxin* produced by *Corynebacterium diphtheriae* inhibits protein synthesis and causes necrosis of epithelium, heart muscle, kidney and nerve tissues;
- *tetanus toxin* produced by *Clostridium tetani* increases reflex excitability in neurones of the spinal cord by blocking release of an inhibitory mediator in motor neurone synapses;
- *botulinus toxin* produced by *Cl. botulinum* affects motor neurones, blocking release of acetylcholine at synapses and neuromuscular junctions and resulting in motor paralyses (e.g. dysphagia, respiratory arrest);
- *staphylococcal enterotoxin* produced by specific strains of *S. aureus* acts on the central nervous system resulting in severe vomiting within hours;
- *clostridial toxins* produced by *Cl. perfringens* cause tissue necrosis and haemolysis;
- *cholera toxin* produced by *Vibrio cholera* binds to ganglioside receptors of the small intestine epithelial cells and results in increased adenylate cyclase activity with massive hypersecretion of chloride and water into the lumen.

CHAPTER 24

Immunology of Infection

The human body protects itself against infection by several mechanisms:
- *non-specific or innate immune defences* represent the initial defences encountered by an organism when it comes into contact with the host;
- *specific or adaptive mechanisms* are designed to act when non-specific defences are breached; these involve the body recognizing specific components of the invading organism (antigens) and the producing of immune factors (antibodies; T-lymphocytes) and finally commits the antigenic detail to immunological memory.

On first encountering a pathogen the specific immune system is slow to react and the body is heavily reliant upon non-specific immune defences. However, subsequent exposure to the same organism elicits a much more rapid response ('booster' response) from the specific immune system; the host has acquired immunity.

The immunology of infectious disease is complex and to help understand it, it is necessary to describe the different systems involved in host defence as separate entities ('non-specific' and 'specific'; 'antibody-dependent' and 'cell-mediated immunity'). However, these components are interdependent and often act synergistically.

NON-SPECIFIC IMMUNE DEFENCES

Non-specific defences can be divided into:
- external biochemical and mechanical barriers;
- phagocytic cells;
- soluble factors (e.g. complement).

External biochemical and mechanical barriers

The importance of these external barriers is best illustrated by considering infections that result from their reduction (Table 24.1).

Non-specific immune defences

IMPORTANCE OF PHYSICAL DEFENCES

Site of entry	Defence mechanism	Condition of defence reduction	Consequence
Skin	Dry and acidic (fatty acids); continuous shedding of normal flora	Burns, wounds, exzema, ulcers	Colonization and infection by *Staphylococcus aureus* and *Pseudomonas* species
Alimentary tract	Acid in stomach Motility Mucous coating Normal flora IgA antibodies	Achlorhydria Antibiotic treatment Ileus (post-surgery)	Oral candidiasis Overgrowth of coliforms in upper gastrointestinal tract, resulting in vomiting and in aspiration pneumonia
Respiratory tract	Nasal filters Mucous trapping Ciliary action Coughing IgA antibodies	Cystic fibrosis Smoking Post-surgery Intubation	Chronic lung infection Post-operative pneumonia
Urinary tract	Outward flow Complete emptying bladder Mucous coating IgA antibodies	Vesico-ureteric reflux Outflow obstruction with residual urine in bladder	Recurrent and chronic urinary tract infection
Genital tract (female)	Acid production by lactobacilli (normal flora) IgA antibodies	Pregnancy and oral contraceptives (reduced acidity) Antibiotic therapy	Vaginal candidiasis

Table 24.1 The importance of physical defences.

Phagocytes

The principal phagocytic cells of the body are macrophages and polymorphonuclear leucocytes (polymorphs or neutrophils).

MACROPHAGES

Macrophages are part of the mononuclear phagocytic cell lineage. Promonocytes in the bone marrow mature into macrophages, and migrate to various body tissues, particularly lymph nodes and spleen, and submucosal tissues. They are often given specific names relating to each tissue (e.g. liver, Kupffer's cells; lung, alveolar macrophages) and have various functions including phagocytosis and killing of organisms, and the synthesis of complement components and cytokines (see below).

POLYMORPHS

Polymorphs originate from precursor cells in the bone marrow, and are found primarily in the circulation, only appearing in tissues as part of an inflammatory response. The majority of circulating polymorphs are sequestrated in the spleen and capillary beds, and are recruited in response to infection. The appearance of polymorphs in the blood results in the increase in peripheral white cell count associated with infection. The primary function of polymorphs is phagocytosis and intracellular killing of microorganisms.

Phagocytosis and intracellular killing

Phagocytosis is the process whereby phagocytic cells engulf particulate matter such as bacteria; it involves several stages (Fig. 24.1).

Chemotaxis. Phagocytes are attracted to the site of inflammation (chemotaxis) in response to a variety of soluble stimuli (chemotaxins), including bacterial products and inflammatory mediators produced by the host (e.g. the complement components C3a and C5a, see below).

Attachment. Bacteria attach to the phagocytic cell membrane by non-specific mechanisms dependent on the physiochemical properties of the bacterial cell surface.

Engulfment. Bacteria are engulfed by the phagocytic cells in a manner similar to that of amoebae engulfing particles, with small pseudopodia developing around the bacterial cell which eventually coalesce and internalize the organism in a 'phagosome'.

Intracellular killing. There are two principal mechanisms whereby phagocytic cells kill internalized bacteria:
1 *oxygen-dependent killing*: toxic oxygen radicals (hydrogen peroxide, superoxide ions) produced by the phagocyte cell membrane are secreted into the phagosome; together with the lysosomal enzyme, myeloperoxidase, and halide ions, they attack the bacterial cell wall;
2 *lysosomal killing*: the phagosome fuses with lysosomes containing enzymes that degrade the engulfed organism.

Non-specific immune defences | 161

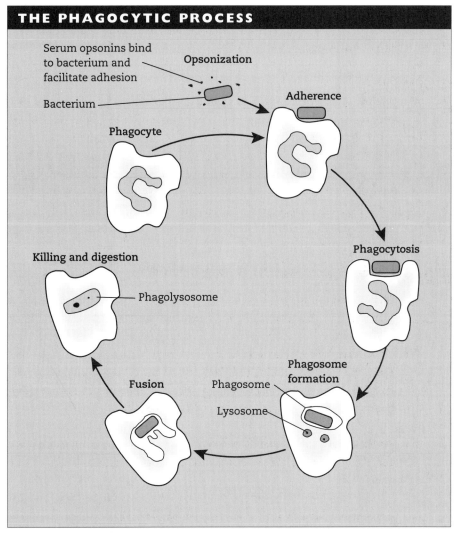

Fig. 24.1 The phagocytic process.

The complement system

The complement system consists of a series of serum proteins which activate each other in a cascade, similar to the clotting system (Fig. 24.2).

COMPLEMENT ACTIVATION

The principal activator is antibody bound to antigen; other activators include bacterial lipopolysaccharide and peptidoglycan.

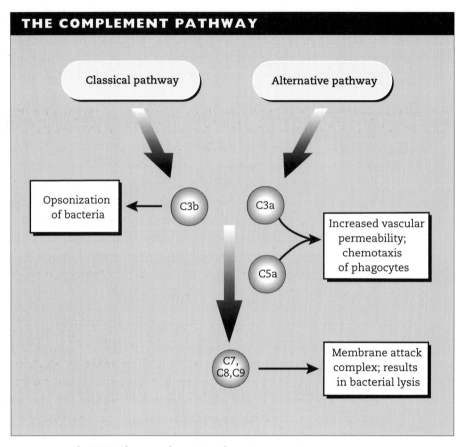

Fig. 24.2 The complement pathway.

COMPLEMENT PATHWAYS

There are two complement pathways, the classical pathway and the alternative pathway; both generate immunologically active proteins (Fig. 24.2).

• C3a and C5a increase vascular permeability and act as chemotaxins, attracting phagocytes to the site of infection.

• C3b binds to the surface of bacteria and mediates immune adherence of bacteria to the phagocyte via the C3b receptor ('opsonization').

• C7/C8/C9 bind to the surface of the bacterial cell resulting in lysis, particularly of Gram-negative organisms.

CONTROL OF COMPLEMENT

As with any cascade system, feedback loops and specific degrading enzymes are important in controlling the production and removal of immunoactive complement components.

COMPLEMENT AND DISEASE

Absence of complement components or enzymes involved in feedback mechanisms can lead to a spectrum of disease ranging from recurrent infection to immunopathological conditions such as systemic lupus.

SPECIFIC DEFENCE MECHANISMS

Most pathogenic organisms have developed strategies (e.g. capsules, toxins) for avoiding the non-specific immune defences; these are often referred to as 'virulence' factors and were discussed in Chapter 23. In response, the host has developed a specific immune system which is based principally on the recognition of foreign proteins, carbohydrates and other molecules (antigens) by specific immunoglobulin (antibodies) and cells (T-lymphocytes). These neutralize the invading organism or its products (e.g. toxins) and crucially support the non-specific immune system to interfere with the pathogen.

Antigens

The term antigen describes particular microbial molecules to which specific antibodies bind, the system being comparable to a lock and key. Antigens on bacterial cell surfaces are often carbohydrates (cell wall and capsular antigens). Protein antigens include the nucleoproteins and glycoproteins of viruses and bacterial exotoxins.

Antibodies

Human antibodies (immunoglobulins) can be classified into five major classes, IgG, IgA, IgM, IgD and IgE, which differ in size and biological functions (Table 24.2). All have the same basic unit structure (Fig. 24.3).

The Fab (fragment antigen binding) portion of antibody is the recognition site that binds to complementary antigen. The Fc site of antibody binds to Fc receptors present on phagocytes, facilitating phagocytosis (opsonization). The Fc receptor can also activate complement by binding to the C3b complement component.

Lymphocytes

Lymphocytes are derived from stem cells which originate in thymus and bone marrow. They are found in lymph nodes and spleen. Lymphocytes are primarily divided into *B-cells* (so-called because they mature in the bone

ANTIBODY CLASSES

Antibody class	Approximate molecular weight	Composition	Comments
IgG	160 000	1 polypeptide unit	80% of total immunoglobulin 4 major subclasses (1–4) Crosses placenta
IgM	900 000	5 polypeptide units	First antibody to appear following infection Does not cross placenta
IgA	160 000–320 000	1 or 2 polypeptide units	Principal mucosal antibody; important in host defence at mucosal surfaces Present in breast milk
IgE	200 000	1 polypeptide unit	Key role in hypersensitivity reactions (allergy) Important in defence against parasitic infections
IgD	200 000	1 polypeptide unit	Mainly found on surface of B-lymphocytes

Table 24.2 Antibody classes: composition and function.

marrow) and *T-cells* (which mature in the thymus). Lymphocytes can be further subdivided according to molecules on their surface; these molecules are known as cluster determinants (CD). CD surface markers are important in the subdivision of T-cells.

B-LYMPHOCYTES

B-lymphocytes express immunoglobulin on their surfaces. During differentiation each B-lymphocyte becomes committed to the expression of an immunoglobulin specific for a particular antigen; the B-cell is then said to be immunocompetent. Foreign antigen (e.g. an invading bacteria) is recognized by lymphocytes expressing the complementary antibody; this leads to clonal proliferation of the original cell resulting in a large population of B-cells secreting the correct antibody. B-cells actively secreting antibody undergo morphological changes and are termed plasma cells.

The clonal proliferation of B-lymphocytes following initial contact with antigen results not only in the production of large numbers of active plasma cells but also a larger pool of B-lymphocytes expressing the particular immunoglobulin. These cells are known as memory cells and are responsible for the more rapid antibody response that occurs when the

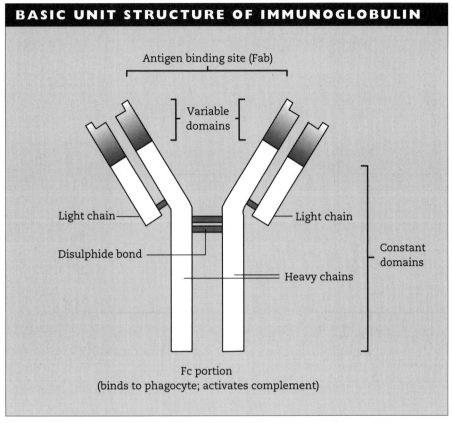

Fig. 24.3 Basic unit structure of immunoglobulin. Two identical heavy polypeptide chains and two identical light polypeptide chains are linked by disulphide bonds. The domains at the Fab portion are variable to reflect binding to different antigens.

T-LYMPHOCYTE SUBTYPES AND FUNCTIONS

Subtype	Function
CD4 (T-helper cells)	Stimulate B-lymphocytes to synthesize antibody Cytokine production
CD8 (T-suppressor cells)	Inhibit T-helper and B-lymphocytes Control of immune response
CD21/CD35	Complement synthesis
CD8 (cytotoxic T-cells)	Recognition of microbial antigens in complexes with cell surface

Table 24.3 T-lymphocyte subtypes and functions.

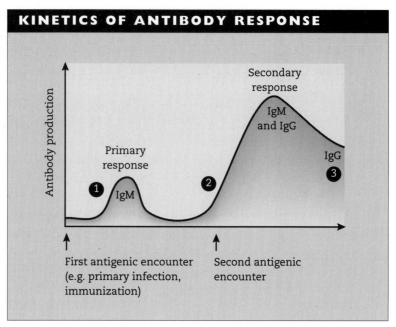

Fig. 24.4 Kinetics of antibody response. (1) First contact with an antigen requires recognition and clonal proliferation of specific B-lymphocytes before antibody production (principally IgM). (2) Subsequent contact with the antigen results in more rapid antibody production (IgM + IgG). (3) Serum IgG antibodies remain raised and can be measured to confirm immune status.

host is challenged with the same antigen a second time (Fig. 24.4). Following a second stimulation there is normally a very low, or no IgM response. Such primary and secondary antibody responses are important when considering immunization (Chapter 47).

T-LYMPHOCYTES

T-lymphocytes also have surface receptors that recognize antigens. As with B-cells, T-cells express a single receptor specific for a particular antigen and clonal expansion of individual T-cells occurs in response to an antigen. Subtypes of T-cells with different functions are recognized by CD surface markers (Table 24.3).

T-lymphocytes and macrophages are important in cell-mediated immunity against viruses and bacteria (e.g. mycobacteria) that can survive phagocyte and antibody/phagocyte defences. Infected phagocytes are recognized and destroyed by recruitment (via cytokines) of macrophages and cytotoxic T-lymphocytes.

Cytokines

Cytokines are soluble factors involved in intercellular communication and include interleukins (various types), interferons (α, β, γ), tumour necrosis factor (TNF) and the colony stimulating factors (e.g. GM-CSF which stimulates growth of granulocytes and monocytes).

The cytokines are the subject of intensive investigation, and a full understanding of the structure and function of individual factors is not yet elucidated. Functions include the proliferation and maturation of T- and B-cells, the activation of polymorphs, antiviral effects, cytotoxicity, and stimulation of growth of granulocytes and monocytes.

Research into cytokines has not only improved our understanding of the immune response, but has led to the development of novel therapeutic agents (e.g. GM-CSF for the treatment of neutropenia; monoclonal antibody to TNF for the treatment of endotoxic shock; interferons as antiviral agents).

CHAPTER 25

Diagnostic Laboratory Methods

This chapter describes the common techniques used in diagnostic microbiology laboratories. Virus culture and detection are described in Chapter 18.

GENERAL PRINCIPLES FOR SPECIMEN COLLECTION

1 Whenever possible, specimens should be obtained before starting or altering antimicrobial therapy, and be collected into sterile containers by techniques that avoid contamination from the normal flora of the patient, the person obtaining specimens, or the environment.

2 Delays in transportation to the laboratory may reduce viability of some organisms and result in overgrowth of others. Special transport media are available but rapid transport to the laboratory is preferable, particularly for fastidious organisms, e.g. *Neisseria gonorrhoea*.

3 Specimens should be transported in leakproof containers in accordance with safety guidelines.

4 The request form should document the patient's basic clinical details, including age, diagnosis, date of onset of symptoms, date and time of specimen collection, and current antimicrobial therapy. When appropriate, history of travel, suspected contact of infection, immunization, or involvement in an outbreak should be given. This information is essential for the laboratory in selecting which investigations to perform.

LABORATORY DIAGNOSIS OF INFECTION

Samples can be examined by microscopy, culture and various antigen–antibody tests. High risk specimens, possibly containing Category 3 organisms e.g. sputum, from a suspected case of tuberculosis are prepared in special cabinets (Plate 37).

Direct microscopy

Samples can be examined either directly (e.g. urine) or in an emulsified suspension (e.g. faeces). Direct microscopy is normally limited to the following investigations:

- the detection of parasites (helminths, protozoa) by visualization of trophozoites, ova or cysts (Plates 34 and 35): faeces for intestinal parasites; urine for schistosomes (Plate 36); vaginal swabs for trichomonas; blood for malaria;
- direct microscopy of urine and CSF samples in order to assess and enumerate inflammatory cells.

Direct microscopy following staining

GRAM STAIN

The Gram stain is the principal stain used in diagnostic microbiology and allows rapid classification of bacteria into four simple categories: Gram-positive cocci, Gram-positive bacilli, Gram-negative cocci or Gram-negative bacilli (Plates 2, 3, 5, 9–11, 15, 16). Fungi can also be identified by the gram stain. It also helps to determine the presence of inflammatory cells.

In most diagnostic microbiology laboratories, Gram stains are performed routinely on specimens of sputum, pus, cerebrospinal fluid (CSF) and swabs of the genital tract.

> **GRAM STAIN**
>
> 1 Spread specimen thinly on glass microscope slide and heat fix
> 2 Stain with methyl violet (blue)
> 3 Add iodine (this fixes the dye in Gram-positive bacteria only)
> 4 Decolorize with acetone or alcohol (only Gram-negative bacteria are decolorized; Gram-positive remain blue)
> 5 Counter stain; basic fuchsin (stains Gram-negative bacteria red; Gram-positive bacteria remain blue/purple)

ZIEHL–NEELSEN'S (ZN) STAIN

ZN stains are performed on specimens, including sputum, pus and urine when mycobacterial infection is suspected (Plate 23).

OTHER SPECIAL STAINS

A number of other stains are used occasionally in the clinical laboratory; these are referred to in the sections describing specific pathogens.

ZIEHL-NEELSEN'S STAIN

1 Spread specimen thinly on glass microscope slide and heat fix
2 Stain with **HOT** carbol-fuchsin dye (all bacteria are stained)
3 Decolorize with 20% sulphuric acid (only mycobacteria retain dye)
4 Counter stain with methylene blue or malachite green to reveal non-mycobacterial organisms

Immunological methods

Immunological detection methods rely on the direct detection of microbial antigens in specimens with fluorescein-labelled antibodies, and are used in routine diagnostic microbiology for the rapid diagnosis of serious infections (e.g. diagnosis of bacterial meningitis), or where a pathogen is difficult or impossible to culture (e.g. *Pneumocystis carinii*, *Chlamydia trachomatis* (Plate 25)).

Electronmicroscopy

Electronmicroscopy is used to visualize and identify viruses in some specimens (Plate 38), e.g. faecal samples from children with diarrhoea or from patients involved in outbreaks of food poisoning; or blister fluids for the rapid diagnosis of *Herpes simplex* (Plate 39) and *Varicella zoster* infections.

Culture techniques

- Most medically important bacteria and fungi can be cultured on laboratory media (Plates 4, 6–8, 13, 14, 17, 18, 22). Culture remains the most reliable method for confirming diagnosis and optimizing treatment by the performance of antibiotic susceptibility tests.
- Various media (Plate 40) and culture conditions are utilized for different specimens, reflecting the likely pathogens. Culture plates are usually incubated either under aerobic conditions (sometimes with added CO_2) or in an anaerobic atmosphere in specialized cabinets (Plate 41).
- There are two basic types of specimen:
  (a) *specimens from normally sterile sites* (e.g. blood (Plate 42), CSF, joint fluid): these are cultured on enriched media to isolate most organisms as any isolate is likely to be significant;
  (b) *specimens from sites with a normal flora* (e.g. upper respiratory tract specimens, faeces, genital specimens): are cultured on selective media designed to suppress normal endogenous flora but allowing likely pathogens to grow.

Laboratory diagnosis of infection

- Sputum and urine specimens are sometimes described as '*clean contaminated*', as normally both should be sterile, but may become contaminated during sampling: sputum by upper respiratory tract flora, urine by perineal flora. Some 'contaminating organisms' may also cause infection (e.g. pneumococci from the pharynx may contaminate sputum but are also an important lower respiratory tract pathogen). Quantitative culture is performed to help distinguish contamination (low numbers of bacteria/ several different bacterial species) from infection (high numbers of single bacterial species). This approach is particularly important in the diagnosis of urinary tract infection (Chapter 36).

Microbial identification

Techniques to identify bacteria and fungi include:
- growth characteristics (e.g. aerobic or anaerobic; Plate 41);
- colonial morphology (e.g. size, haemolysis, colour);
- microscopy of organism with or without staining (wet preparations of fungal colonies are particularly important in speciation of medically important fungi);
- antigenic characteristics (e.g. Lancefield grouping of β-haemolytic streptococci; speciation of *Salmonella*);
- biochemical tests (e.g. fermentation of sugars; Plates 43 and 44);
- growth factor requirements (e.g. X and V dependency of *Haemophilus influenzae*; Plate 19);
- production of toxins (e.g. Nagler test for *Clostridium perfringens* (Plate 12; see Chapter 7); Elek test for *Corynebacterium diphtheriae* (Fig. 25.1; see Chapter 8)).

Identification to species level is normally satisfactory in a diagnostic laboratory. Further characterization of individual isolates ('typing') may be important when defining outbreaks of cross-infection; techniques include phage typing, plasmid analysis and characterization of chromosomal DNA.

Antibiotic sensitivity testing

Significant isolates are tested for susceptibility to different antimicrobial agents by various techniques (e.g. disc testing (Plate 45)); these are described in Chapter 26.

Antigen–antibody interactions

Antigen–antibody interactions have developed into an important part of laboratory diagnosis. The basic principle is that for each microbial antigen

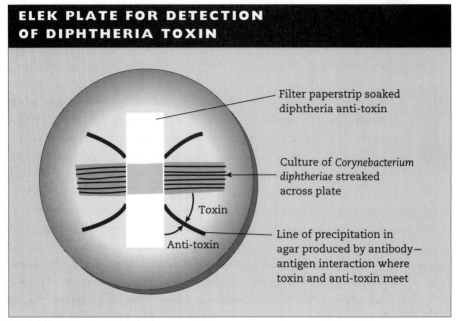

Fig. 25.1 Elek plate for detection of diphtheria toxin.

there is a specific antibody. Antigen–antibody reactions can be used in developing tests to detect *specific organisms* or *specific antibody*; the first is termed 'typing' the latter 'serology'. The antigen–antibody reaction created must be made visible and can be achieved by various techniques.

Precipitation reactions. These include the Elek test for diphtheria toxin (Fig. 25.1).

Agglutination reactions. An example of this is *Salmonella* speciation; known antisera to the various *Salmonella* species are mixed individually with suspensions of an unknown *Salmonella* isolate; an agglutination reaction (visible clumping of the bacteria) with a particular antisera indicates the serotype.

Co-agglutination reactions. Known antibody *or* antigen is fixed to an inert particle (e.g. latex beads). When antigen–antibody reaction occurs, the beads agglutinate; examples include:
- Lancefield grouping of streptococci;
- rapid diagnosis of bacterial meningitis by detection of bacterial antigen in CSF;

- identification of *Neisseria gonorrhoea* (Plate 46);
- identification of *S. aureus*.

Antibody labelling
- *Antigen detection* (Table 25.1): fluorescein-labelled specific antibodies may be used to detect some pathogens directly in specimens (Fig. 25.2); the specimen is viewed by fluorescence microscopy. In positive samples fluorescent foci are seen (Fig. 25.2; Plate 25). Specific antibodies may also be labelled with enzymes (enzyme-linked immunosorbent assay (ELISA)) which, following binding, can be visualized by the addition of a substrate which is converted to a coloured product. The end-point may be read by a colorimeter.
- *Antibody detection* (Fig. 25.3): measurement of antibody by labelled antibody techniques requires an additional step. These assays are often carried out in microtitration trays. Individual wells are precoated with known microbial antigen and the patient's serum added. The well is washed to remove non-specific binding and then remaining bound antibody detected with antibody to human immunoglobulin labelled with fluorescein or enzyme (ELISA; Fig. 25.3). Antibodies of the various human antibody

ORGANISM AND ANTIGEN DETECTION

Specimen	Organism	Technique
Respiratory secretions	Influenza viruses Parainfluenza viruses Respiratory syncytial virus Adenoviruses Measles virus *Legionella pneumophila* *Pneumocystis carinii*	Immunofluorescence
	Group A streptococci	ELISA
Urogenital specimens	*Chlamydia trachomatis*	Immunofluorescence or ELISA
Cerebrospinal fluid	*Haemophilus influenzae* *Streptococcus pneumoniae* *Neisseria meningitidis* Group B streptococci	Co-agglutination
Faeces	Rotaviruses Adenoviruses	ELISA

Table 25.1 Direct detection of organisms or their antigens in clinical specimens.

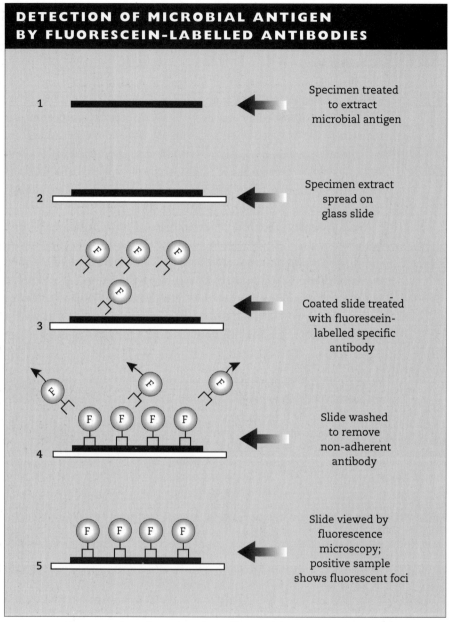

Fig. 25.2 Direct detection of organisms or their antigens by fluorescein-labelled antibodies.

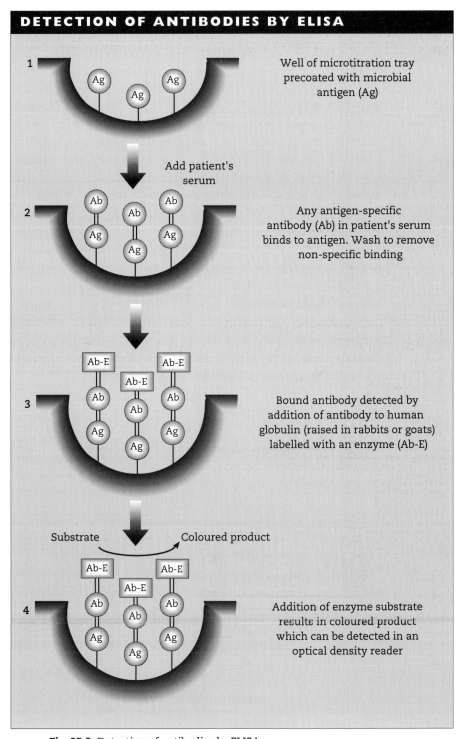

Fig. 25.3 Detection of antibodies by ELISA.

classes, IgG, IgM or IgA can be raised, and each can be detected individually.

Serology
- A variety of serological techniques are used in diagnostic microbiology:
    (a) ELISA and fluorescent labelling are described above; many commercial and automated kits are now available;
    (b) complement fixation test: assay based on the degradation of complement in the presence of antigen–antibody complexes (measured by lack of lysis of haemolysin-erythrocyte complexes, 'activated red blood cells'); gradually being replaced by ELISA;
    (c) agglutination and co-agglutination methods; often used for rapid screening of sera.
- Serology is important for the diagnosis of infections caused by pathogens which are difficult to culture by standard laboratory methods, e.g. *Mycoplasma pneumoniae*, syphilis, hepatitis A, B and C, HIV.
- Active infection is diagnosed by either detection of specific IgM or a fourfold increase in IgG antibodies in paired sera (acute and convalescent samples) taken 10–14 days apart.
- Immune status testing (e.g. to test response to hepatitis B or rubella vaccination) is performed by detection of IgG antibodies.

Other diagnostic methods

Gas–liquid chromatography. This can detect the presence of volatile fatty acids produced by anaerobic bacteria in pus samples. It is occasionally used for the rapid detection of anaerobes.

DNA technology. Molecular biology techniques for the detection of microbial nucleic acids in specimens have been developed and are beginning to be introduced into diagnostic microbiology laboratories.

The first assays developed relied on DNA hybridization and lacked sensitivity. Newer assays which multiply the DNA present before detection (polymerase chain reaction) are now available; these have much greater sensitivity and may prove useful for organisms that are difficult or slow to culture (e.g. mycobacteria, chlamydiae; viruses, e.g. HBV, HCV, HIV, HSV, VZV).

CHAPTER 26

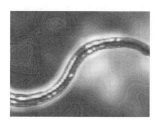

Antibacterial Agents

DEFINITIONS

Antimicrobial agent. Substance with inhibitory properties against microorganisms (includes antibiotics and synthetic compounds) but with minimal effects on mammalian cells ('selective poisons').

Antibiotic. Substance produced by microorganisms which inhibits the growth of (bacteriostatic) or kills (bactericidal, viricidal, fungicidal) other microorganisms. The term 'antibiotic' is often incorrectly used to include all antimicrobial agents, some of which are synthetic, e.g. sulphonamides.

Semi-synthetic antibiotics. Antibiotics chemically altered to improve properties, e.g. stability or spectrum of activity.

Mechanism of action. Antimicrobial agents are divided into groups according to their mode of action. The potential targets for antimicrobial action include the bacterial cell wall, cell membrane, protein synthesis and nucleic acid synthesis (Table 26.1).

INHIBITORS OF CELL-WALL SYNTHESIS

Bacteria, unlike mammalian cells, have a cell wall. The main groups of antimicrobial agents that act selectively on the bacteria cell wall are the β-lactams and glycopeptides.

β-lactam antibiotics

Structure. This is a large group of compounds, all with a beta-lactam ring (Fig. 26.1). The structure of the side-chains and rings attached to the β-lactam ring determines the class of antibiotic and also its properties. Cephalosporins have a six-membered ring, whereas penicillins have a five-membered ring. Many modifications have been made to the penicillins and cephalosporins by addition of structural groups at various sites.

SITES OF ACTION

Site of action	Examples of antimicrobials
Cell wall	Penicillins Cephalosporins Vancomycin
Cell membrane	Polymyxins
Protein synthesis	Aminoglycosides Chloramphenicol Fucidic acid Macrolides Tetracyclines
DNA synthesis	Sulphonamide Trimethoprim Quinolones
RNA synthesis	Rifampicin

Table 26.1 Sites of action of different antimicrobial agents.

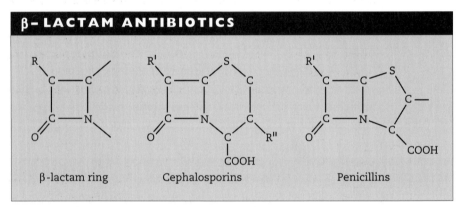

Fig. 26.1 Structure of β-lactam antibiotics.

Action. They interfere with the cross-linking of the cell-wall polymer peptidoglycan by inhibiting carboxypeptidases and transpeptidases (penicillin binding proteins (PBPs)) reactions that form a link between the building blocks of peptidoglycan, *N*-acetylmuramic acid and *N*-acetylglucosamine; this weakens the cell wall and results in rupture (lysis) of the microorganism (Plate 48). The antibiotics within this group include the penicillins (e.g. benzylpenicillin) and the cephalosporins (e.g. cefuroxime).

PENICILLINS

The penicillins comprise a large group of mainly semisynthetic compounds based on benzylpenicillin. They can be divided into the following groups:

1 benzylpenicillin, active against many streptococci;
2 orally absorbed penicillins, similar to benzylpenicillin (e.g. penicillin V);
3 penicillins resistant to staphylococcal β-lactamase (e.g. flucloxacillin);
4 extended spectrum penicillins, with activity against streptococci and many coliforms (e.g. ampicillin);
5 penicillins with activity against *Pseudomonas aeruginosa* (e.g. piperacillin).

- *Pharmacokinetics.* The penicillins have a wide distribution in the body. Excretion is mainly renal.
- *Toxicity.* Hypersensitivity reactions include anaphylaxis and urticarial skin rashes (maculopapular); 10% of penicillin-allergic patients are also allergic to cephalosporins.
- *Antibacterial resistance*

    (a) Alteration in target site, e.g. methicillin-resistant *Staphylococcus aureus* can produce a PBP with low affinity for β-lactams and therefore continues to function in the presence of the antibiotic.

    (b) β-lactamase production: this bacterial enzyme hydrolyses the β-lactam ring; some β-lactams are unaffected by β-lactamases (i.e. are β-lactamase-stable).

    (c) Cell membrane alterations, reducing uptake in Gram-negative bacteria.

Examples of penicillins

Benzylpenicillin

- *Pharmacokinetics.* Unstable in acid and destroyed in the stomach; can be given intravenously (i.v.) or intramuscularly (i.m.). Widely distributed in the body. Excretion is predominantly renal.
- *Antimicrobial spectrum of activity.* Effective mainly on Gram-positive organisms and Gram-negative cocci, including streptococci, neisseriae and *Treponema pallidum*.
- *Clinical applications.* Streptococcal infections, pneumococcal pneumonia, meningococcal meningitis, gonorrhoea and syphilis.

Flucloxacillin

- *Pharmacokinetics.* This is a semisynthetic penicillin. It is well absorbed orally; part metabolized in the liver, part excreted in urine.
- *Antimicrobial spectrum of activity.* Flucloxacillin is active against both

β-lactamase-positive and -negative strains of *S. aureus*, *Streptococcus pyogenes*.
- *Clinical applications.* *S. aureus* infections, including abscesses pneumonia, endocarditis and septicaemia.

Ampicillin
- *Pharmacokinetics.* A semisynthetic penicillin; acid-stable and well absorbed orally. Excretion is predominantly renal.
- *Antimicrobial spectrum of activity.* Similar to benzylpenicillin, but more active against *Enterococcus faecalis*, *Haemophilus influenzae* and some Gram-negative aerobic bacilli. Activity against many β-lactamase-producing bacteria can be enhanced by the co-administration of β-lactamase inhibitors, such as clavulanic acid.
- *Clinical applications.* Urinary and respiratory tract infections caused by *H. influenzae*, some coliforms; *Listeria monocytogenes* meningitis; infections with *E. faecalis*.

Piperacillin
- *Pharmacokinetics.* A semi-synthetic penicillin, piperacillin is not absorbed orally. The principal route of excretion is via the kidneys.
- *Antimicrobial spectrum of activity.* Piperacillin is moderately active against many Gram-positive organisms, neisseriae, *H. influenzae*. It has wider activity than ampicillin against coliforms and is active against *Pseudomonas aeruginosa*.
- *Clinical application.* Serious Gram-negative sepsis.

CEPHALOSPORINS

The cephalosporins are a large group of antimicrobial agents based on cephalosporin C with a structure similar to the penicillins (Fig. 26.1). They are classified as:
1 first generation: early compounds;
2 second generation: compounds resistant to β-lactamases;
3 third generation: compounds both resistant to β-lactamases and with an increased spectrum of activity.
- *Antimicrobial spectrum of activity.* Variable but generally include *S. aureus* (including β-lactamase-positive strains), streptococci, many Gram-negative species, including neisseriae, *Haemophilus* and coliforms.
- *Toxicity/side-effects.* Hypersensitivity with rashes, usually maculopapular.
- *Resistance.* Similar to the penicillins.

Examples of cephalosporins

Cefuroxime
- *Pharmacokinetics.* Available in oral and parenteral forms. Wide distribution in the body with renal excretion.
- *Antimicrobial spectrum of activity.* S. aureus, most streptococci, coliforms. P. aeruginosa, enterococci and anaerobes are resistant.
- *Clinical applications.* Urinary tract infections, soft tissue and chest infections, septicaemia.

Ceftazidime
- *Pharmacokinetics.* Not absorbed orally. Given i.m. or i.v. and is distributed into many body tissues and fluids. Excretion is exclusively renal.
- *Antimicrobial spectrum of activity.* Less activity against Gram-positive bacteria but wide activity against Gram-negative aerobic bacteria including P. aeruginosa.
- *Clinical application.* Severe Gram-negative aerobic bacillary infections, (e.g. septicaemia, peritonitis) due to coliforms and P. aeruginosa.

Glycopeptides

The glycopeptide antibiotics, vancomycin and teicoplanin, interfere with peptidoglycan assembly, but at a different site than β-lactams, resulting in cell lysis and death.
- *Pharmacokinetics.* Not absorbed from the gastrointestinal tract. Glycopeptides are given i.v. and are widely distributed; largely excreted by the renal route; poor penetration into CSF.
- *Mechanism of action.* Interfere with cell-wall synthesis by binding to the pentapeptide chain, preventing incorporation of new subunits into the cell wall.
- *Antimicrobial spectrum of activity.* Glycopeptides are bactericidal against Gram-positive bacteria, including staphylococci. They have no activity against Gram-negative bacteria.
- *Toxicity.* Vancomycin requires monitoring of serum levels because of ototoxicity and nephrotoxicity.
- *Resistance.* Occasionally seen in enterococci.
- *Clinical applications.* Severe staphylococcal infections, including endocarditis, peritonitis; infections of prosthetic devices caused by coagulase-negative staphylococci; oral treatment of *Clostridium difficile*-associated pseudomembranous colitis.

ANTIMICROBIAL AGENTS ACTING ON CELL MEMBRANE

Only a few agents target the microbial membrane. These include polymyxins and the polyene and imidazole antifungals. Polymyxins (e.g. colistin) are similar to detergents and disrupt the membrane, resulting in loss of cytoplasmic content. They are nephrotoxic and only used as topical agents, e.g. in gut sterilization regimens.

INHIBITORS OF PROTEIN SYNTHESIS

This group includes the aminoglycosides, macrolides, chloramphenicol and tetracycline. These target the difference between the human and bacterial ribosomes, the latter having 30 and 50 S subunits:
- aminoglycosides (e.g. gentamicin and tobramycin) inhibit protein synthesis by blocking the formation of the initiation complex after messenger RNA has been utilized;
- macrolides (e.g. erythromycin) bind to the 50 S subunit of ribosomal RNA and inhibit the formation of initiation complexes;
- chloramphenicol binds to the 50 S subunit and interferes with the linkage of amino acids in the peptide chain formation;
- tetracyclines bind to the 30 S subunit, preventing binding of aminoacyl transfer RNA to the acceptor site in the ribosome, thereby inhibiting amino acid chain elongation.

Aminoglycosides

- *Examples*. Gentamicin, tobramycin, netilmicin, amikacin, streptomycin.
- *Pharmacokinetics*. Poorly absorbed from the gut; they have poor penetration into tissue and fluids; excretion is almost entirely by the kidneys. Serum levels require monitoring with careful dosage adjustment particularly in renal failure.
- *Antimicrobial spectrum of activity*. Active against staphylococci and Gram-negative aerobic bacilli including *Pseudomonas*; they have no activity against anaerobes. They are bactericidal, showing synergy with β-lactams.
- *Toxicity/side-effects*. Hypersensitivity, ototoxicity and nephrotoxicity.
- *Resistance*. Mechanisms include changes in ribosomal binding of the drug, decreased permeability, and inactivation by aminoglycoside-modifying enzymes.
- *Clinical applications*. Severe sepsis due to coliforms and other Gram-negative aerobic bacilli; in combination with β-lactams for treatment of

Gram-positive infections, e.g. staphylococcal and streptococcal endocarditis.

Macrolides and lincosamides

MACROLIDES

- *Examples.* Erythromycin and azithromycin.
- *Pharmacokinetics.* Absorbed following oral administration; also given i.v. They are well distributed and penetrate into phagocytic cells. Excretion is mainly in the bile.
- *Antimicrobial spectrum of activity.* Most Gram-positive organisms, *Haemophilus, Bordetella, Neisseria* and some anaerobes. Many coliforms are resistant. Active against chlamydiae, rickettsiae and mycoplasmas.
- *Toxicity/side-effects.* Gastrointestinal upsets, rashes, hepatic damage (rare).
- *Resistance.* This is by alteration of the RNA target. (*Note:* there is cross-resistance with clindamycin.)
- *Clinical applications.* Streptococcal and staphylococcal soft tissue and bone infections; respiratory infections caused by *Haemophilus influenzae*, mycoplasmas, legionellae, chlamydiae, campylobacter enteritis.

LINCOSAMIDES

- *Examples.* Lincomycin and clindamycin.
- *Pharmacokinetics.* Usually given orally, but can be given i.v. or i.m. Penetrates well into the bone; metabolized in the liver.
- *Antimicrobial spectrum of activity.* Similar to erythromycin, but more active against anaerobes.
- *Toxicity.* Associated with pseudomembranous colitis, but probably no more than many other antibiotics.
- *Resistance.* See erythromycin.
- *Clinical applications.* They are used in staphylococcal infections, particularly osteomyelitis, as an alternative to flucloxacillin, and in anaerobic infections.

Chloramphenicol

- *Pharmacokinetics.* Rapidly absorbed following oral administration with good penetration into many tissues including CSF. Metabolized in the liver to inactive metabolites, which are excreted via the kidney.
- *Antimicrobial spectrum of activity.* Effective against a wide range of organisms, including Gram-negative and Gram-positive bacteria, chlamydiae, mycoplasmae, rickettsiae.

- *Toxicity/side-effects.* Chloramphenicol has a dose-related, but reversible depressant effect on bone marrow; rarely, irreversible, potentially fatal, marrow aplasia. Toxicity in neonates causes grey baby syndrome.
- *Resistance.* This occurs due to inactivation by an inducible acetylase enzyme; reduced permeability.
- *Clinical applications.* Meningitis, typhoid; topical for eye infections.

Tetracyclines

- *Pharmacokinetics.* Can be given orally or i.v. Penetrate well into body fluids and tissues due to lipid solubility. Excretion is via the kidney and bile duct.
- *Antimicrobial spectrum of activity.* Broad spectrum of activity against many Gram-positive and some Gram-negative bacteria, chlamydiae, mycoplasmae, and rickettsiae.
- *Toxicity/side-effects.* Gastrointestinal intolerance; deposition in developing bones and teeth preclude use in young children and in pregnancy; hypersensitivity occasionally, with skin rashes.
- *Resistance.* Inhibition of transport across cell wall.
- *Clinical application.* Important in treatment of infections caused by mycoplasmae, rickettsiae and chlamydiae.

INHIBITION OF NUCLEIC ACID SYNTHESIS

Several classes of antimicrobial agents act on microbial nucleic acid, including the quinolones, sulphonamides, diaminopyrimidines rifampicin, and nitroimidazoles.

Quinolones

- *Examples.* Nalidixic acid; fluoroquinolones, e.g. ciprofloxacin, ofloxacin and norfloxacin.
- *Mechanism of action.* Inhibit the action of bacterial DNA gyrases (topoisomerases) which are important in 'supercoiling' (folding and unfolding DNA during synthesis).

NALIDIXIC ACID

- *Pharmacokinetics.* Rapidly absorbed following oral administration and almost entirely excreted in urine.
- *Antimicrobial spectrum of activity.* Bacteriostatic against a wide range of

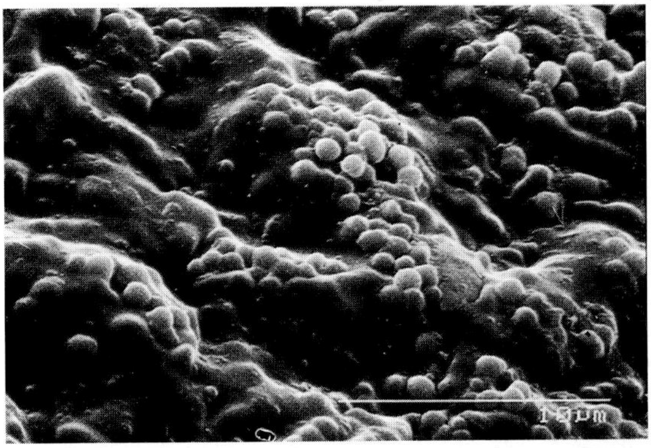

Plate 1 Scanning electronmicrograph showing *Staphylococcus epidermidis* attached to a catheter.

Plate 2 Gram-stain of *Clostridium sporogenes* (showing oval subterminal spores) and a *Clostridium tetani* with a terminal spore.

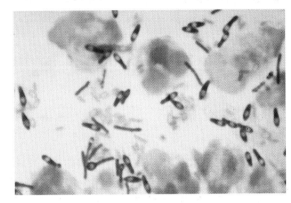

Plate 3 Gram-stain of sputum of patient with *Staphylococcus aureus* pneumonia.

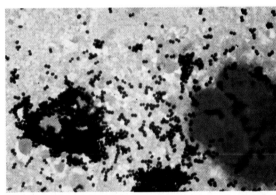

Plate 4 *Staphylococcus aureus* 'golden' colonies on blood agar.

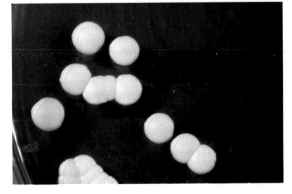

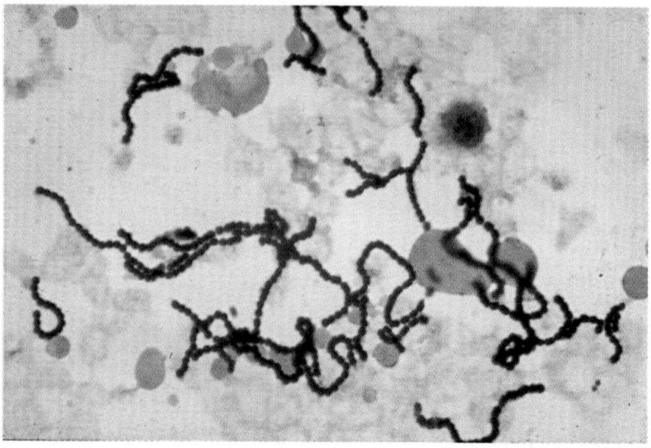

Plate 5 Gram-stain of streptococci isolated from blood, showing long chains.

Plate 6 α-haemolytic streptococci colonies on blood agar.

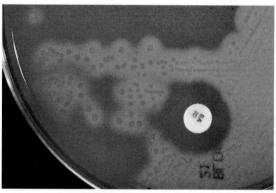

Plate 7 β-haemolytic streptococci colonies on blood agar, sensitive to bacitracin.

Plate 8 Blood agar plate with *Streptococcus pneumoniae* 'draughtsmen' colonies.

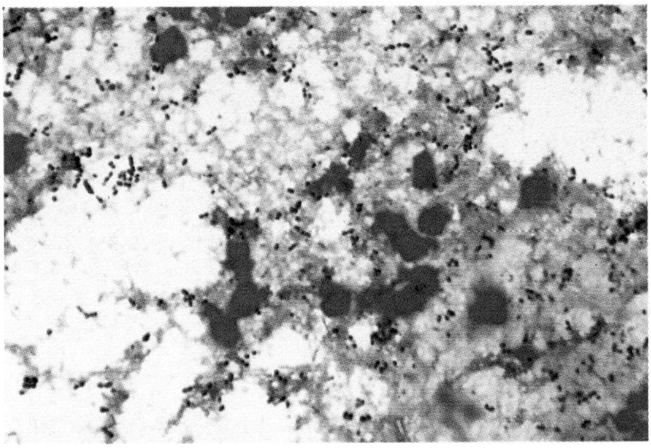

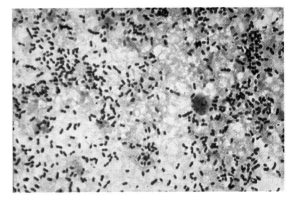

Plate 9 Gram-stain of sputum containing *Streptococcus pneumoniae*.

Plate 10 Gram-stain of cerebrospinal fluid from a patient with overwhelming *Streptococcus pneumoniae* meningitis.

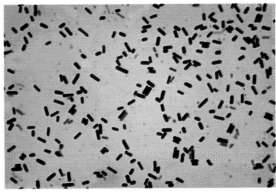

Plate 11 Gram-stain of *Clostridium perfringens*.

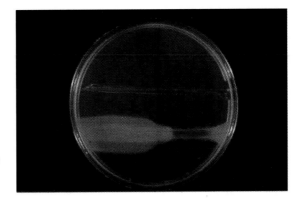

Plate 12 Nagler reaction showing *Clostridium perfringens* (positive) and *Clostridium butyricum* (negative).

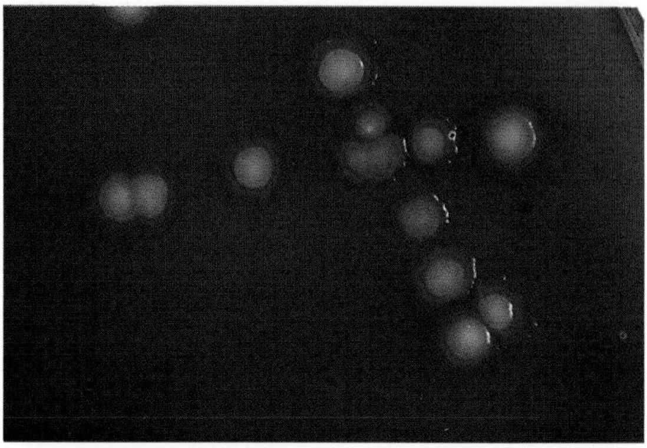

Plate 13 *Clostridium difficile* colonies on blood agar.

Plate 14 Colonies of *Corynebacterium diphtheriae*.

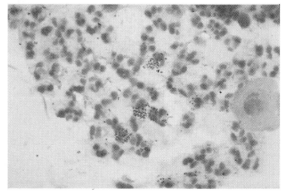

Plate 15 Gram-stain of *Neisseria gonorrhoea* in urethral exudate.

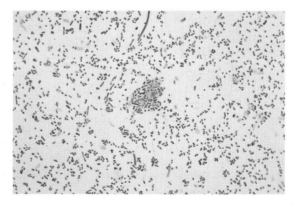

Plate 16 Gram-stain of *Escherichia coli*.

Plate 17 *Escherichia coli* colonies on blood agar.

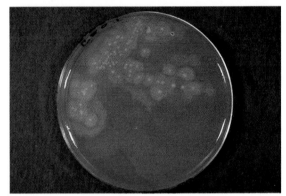

Plate 18 Colonies of Proteus species on blood agar, demonstrating swarming.

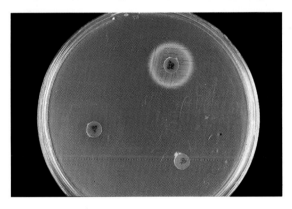

Plate 19 *Haemophilus influenzae* demonstrating requirement for X and V factor for growth. No growth around X or V discs.

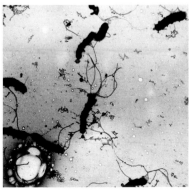

Plate 20 Negative-stain electronmicrograph of *Helicobacter pylori* showing flagella.

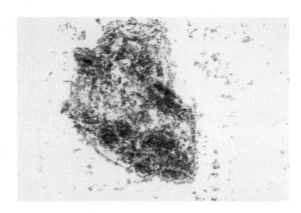

Plate 21 Microscopy of vaginal discharge, showing 'clue cell' with numerous organisms attached to an epithelial cell.

Plate 22 Colonies of *Prevotella melaninogenicus* on blood agar.

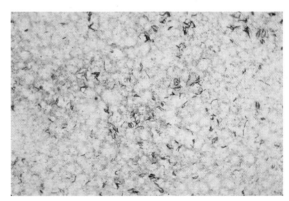

Plate 23 Ziehl–Neelsen stain of *Mycobacterium tuberculosis* showing acid fast bacilli.

Plate 24 Colonies of *Mycobacterium tuberculosis* on Lowenstein Jensen slopes.

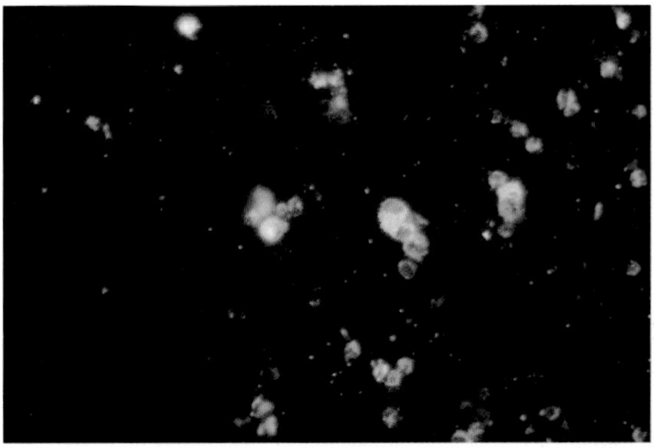

Plate 25 Fluorescence microscopy for chlamydia in smears.

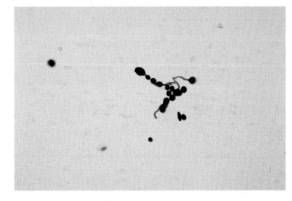

Plate 26 Microscopy of *Candida albicans* showing germ-tube formation.

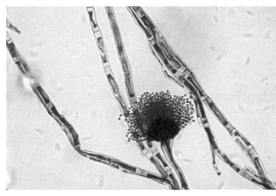

Plate 27 Microscopy of conidiophore of *Aspergillus fumigatus* stained with lactophenol cotton blue.

Plate 28 Colonies of *Candida albicans* on blood agar.

Plate 29 *Aspergillus fumigatus* (left) and *Aspergillus niger* (right) growing on Sabouraud's agar.

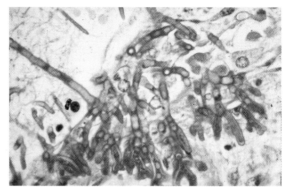

Plate 30 Invasive aspergillus hyphae in brain.

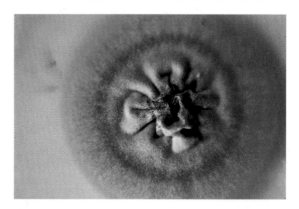

Plate 31 Colony of *Epidermophyton floccosum*.

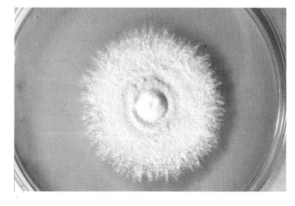

Plate 32 Colony of *Trichophyton mentagrophytes* on malt agar.

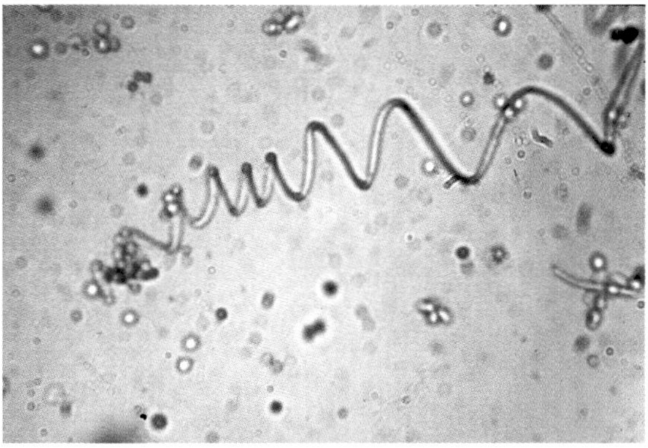

Plate 33 Microscopy of spiral hyphae of *Trichophyton mentagrophytes*.

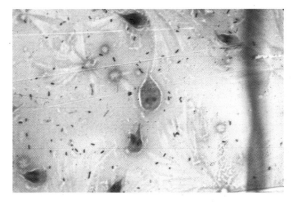

Plate 34 Microscopy of *Giardia lamblia* trophozoites following Giemsa staining.

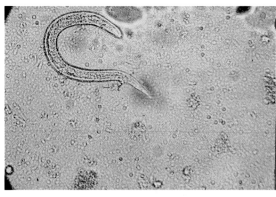

Plate 35 Microscopy of *Strongyloides stercoralis* larva in iodine preparation.

Plate 36 *Schistosoma mansoni* ova.

Plate 37 Class 1 safety cabinet for dealing with dangerous pathogens.

Plate 38 Electronmicroscope.

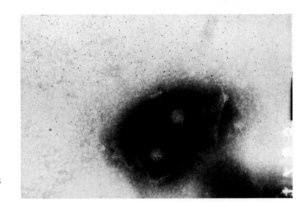

Plate 39 Herpes simplex virus seen after negative staining.

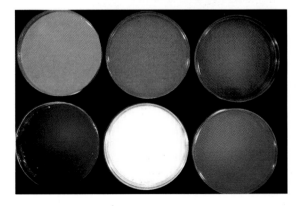

Plate 40 Illustration of some different agar plates used for identifying bacteria.

Plate 41 Anaerobic cabinet.

Plate 43 Method for identifying *Neisseria gonorrhoea* by biochemical reactions.

Plate 42 Blood cultures.

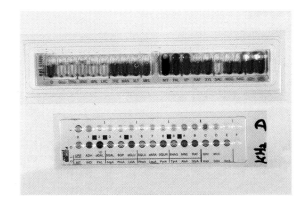

Plate 44 Commercial biochemical-based identification system.

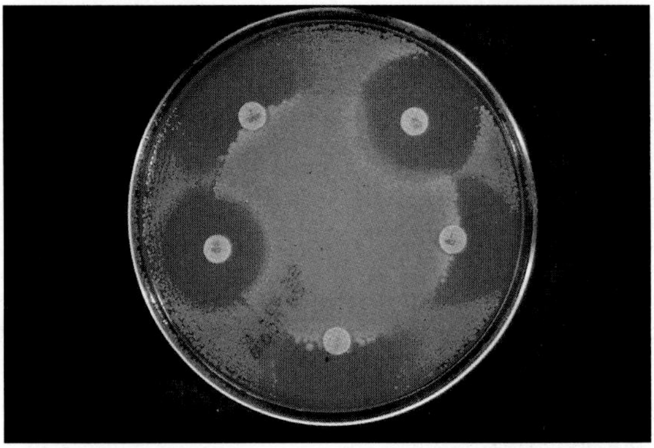

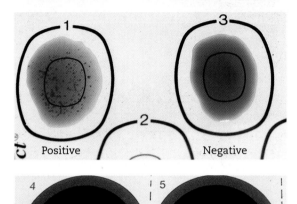

Plate 45 Antibiotic sensitivities of a staphylococcus by Stokes method. Zones of inhibition around the antibiotic impregnated discs for the test organism (inner circle) and control (outer circle) are compared.

Plate 46 Identification of *Neisseria gonorrhoea* by coagglutination method.

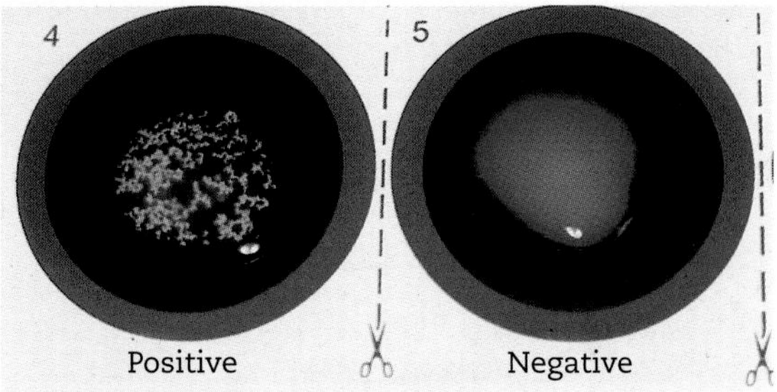

Plate 47 Latex agglutination test for *Staphylococcus aureus*.

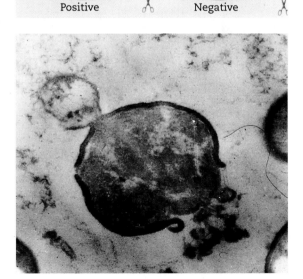

Plate 48 Electronmicrograph illustrating effect of penicillin breaking cell wall of a susceptible bacterium.

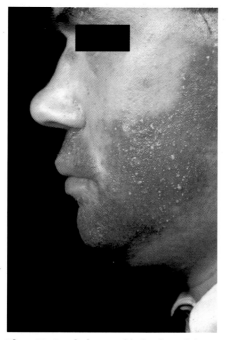

Plate 49 Staphylococcal infection of the face.

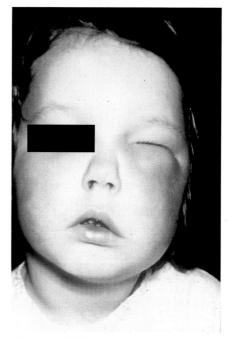

Plate 50 Cellulitis of orbit with associated erythema.

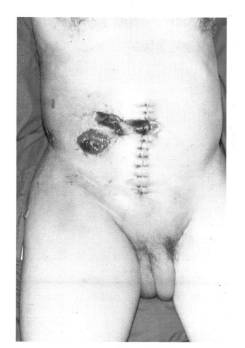

Plate 51 Post-operative cellulitis.

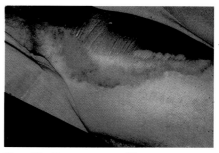

Plate 52 Gas gangrene of leg.

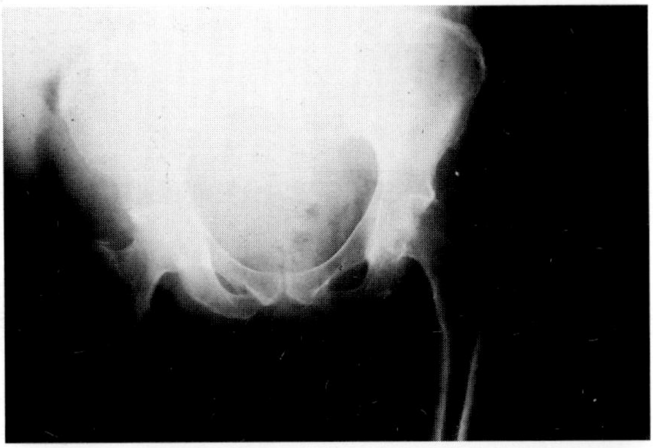

Plate 53 X-ray showing osteomyelitis of left hip.

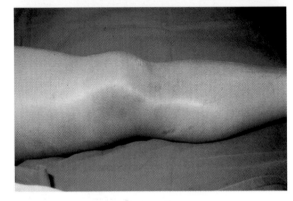

Plate 54 Septic arthritis of knee.

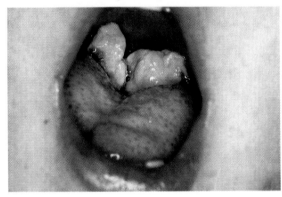

Plate 55 Case of diphtheria showing false membrane on tonsils.

Plate 56 Chest X-ray showing multiple mycyetomas of aspergillus.

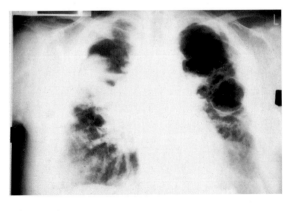

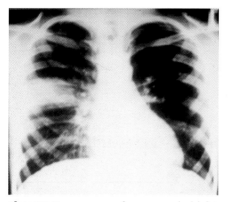

Plate 57 Pneumococcal pneumonia (right middle lobe).

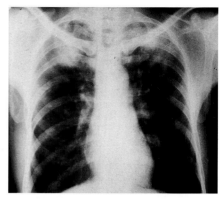

Plate 58 Pulmonary tuberculosis (right apex).

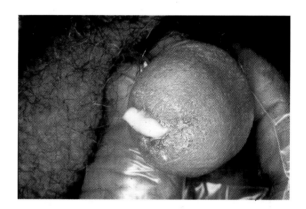

Plate 59 Gonococcal urethral discharge.

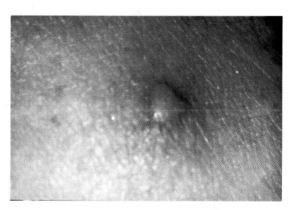

Plate 60 Gonococcal skin lesion.

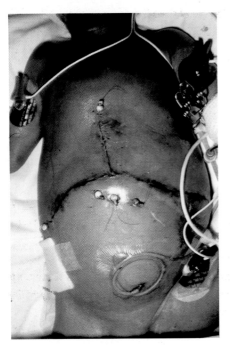

Plate 61 Liver transplant patient with numerous intravascular and drainage catheters.

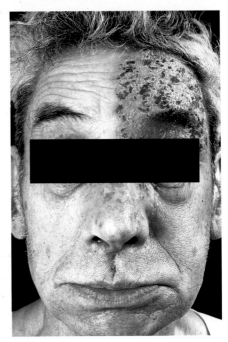

Plate 63 Shingles on face.

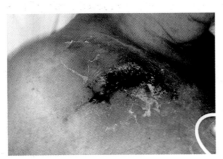

Plate 62 Catheter-related sepsis at insertion site of device.

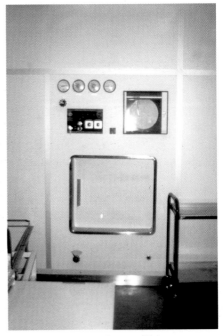

Plate 64 Autoclave for sterilizing instruments.

Enterobacteriaceae; no activity against *Pseudomonas aeruginosa* or Gram-positive organisms.
- *Toxicity.* Gastrointestinal and neurological disturbances.
- *Resistance.* Changes in DNA gyrases result in reduced affinity for nalidixic acid.
- *Clinical application.* Simple urinary tract infections.

FLUOROQUINOLONES

- *Pharmacokinetics.* Generally good absorption following oral administration; penetrate well into body tissues and fluids. Eliminated by renal excretion and liver metabolism, with some excretion in bile.
- *Antimicrobial activity.* Fluoroquinolones have a wider spectrum of activity than nalidixic acid; includes *Pseudomonas*, Enterobacteriaceae, legionellae, chlamydiae, mycoplasmas, rickettsiae. Some activity against Gram-positive bacteria but little effect on anaerobes.
- *Toxicity and side-effects.* Gastrointestinal disturbances, photosensitivity, rashes, neurological disturbances, including seizures (rare).
- *Clinical applications.* Urinary tract infections, gonorrhoea, respiratory and other infections due to Gram-negative aerobic bacilli, including *Pseudomonas aeruginosa*. Also effective for enteritis caused by shigella and salmonella.

Sulphonamides

- *Examples.* Sulphadiazine; sulphadimidine.
- *Mechanism of action.* Act on folic acid synthesis as competitive antagonists of para-aminobenzoic acid, inhibiting purine and thymidine synthesis (Fig. 26.2). Trimethoprim also acts on the folic acid synthesis in the next step in the metabolic pathway.
- *Pharmacokinetics.* Well absorbed following oral administration. Distributed throughout the body and excreted mainly in the urine.
- *Antimicrobial spectrum of activity.* Broad spectrum of activity, including streptococci, neisseriae, *H. influenzae*.
- *Toxicity.* Hypersensitivity reactions, causing renal damage, rash; rarely Stevens–Johnson syndrome (a serious form of erythema multiforme); bone marrow failure.
- *Resistance.* Many Enterobacteriaceae are now resistant; due to production of an altered dihydropteroate synthetase with decreased affinity for sulphonamide.
- *Clinical applications.* Limited by resistance but include nocardiasis and in

combination with trimethoprim (co-trimoxazole) for the prevention and treatment of *Pneumocystis carinii* pneumonia.

Diaminopyrimidines

TRIMETHOPRIM

- *Mechanisms of action.* Prevents synthesis of tetrahydrofolic acid by inhibiting dihydrofolate reductase (Fig. 26.2).
- *Pharmacokinetics.* Rapidly absorbed following oral administration. Excretion is via urine.

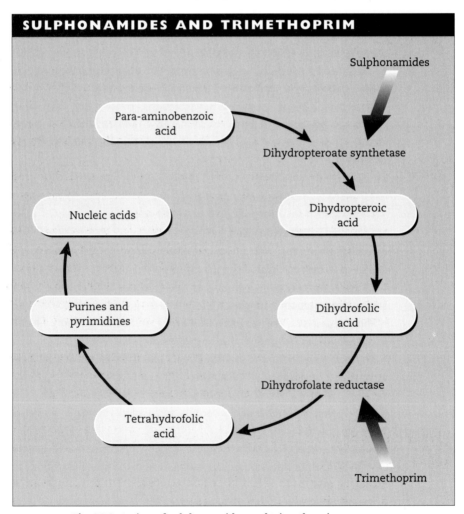

Fig. 26.2 Action of sulphonamides and trimethoprim.

- *Antimicrobial spectrum of activity.* Broad spectrum of activity against many Gram-positive and Gram-negative bacteria. *Pseudomonas* species are resistant.
- *Toxicity.* Folate deficiency.
- *Resistance.* Due mainly to modification of the target enzyme, dihydrofolate reductase.
- *Clinical applications.* Urinary tract and respiratory infections.
- *Co-trimoxazole* (trimethoprim combined with sulphamethoxazole). This was prescribed commonly in general practice for the treatment of urinary tract and respiratory infections; however it is now recognized that trimethoprim alone is probably just as effective and does not have the toxicity problems associated with sulphonamides.

Rifamycins

RIFAMPICIN

- *Mechanism of action.* Binds to RNA polymerase and blocks synthesis of mRNA.
- *Pharmacokinetics.* Well absorbed following oral administration; widely distributed in the body, including CSF. Metabolized in the liver and elimination is primarily by biliary secretion.
- *Antimicrobial spectrum of activity.* Active against many staphylococci, streptococci, *H. influenzae*, neisseriae, legionellae and mycobacteria. Enterobacteriaceae and *P. aeruginosa* are resistant.
- *Toxicity.* Adverse reactions include skin rashes and transient liver function abnormalities; a rare cause of hepatic failure. Potent inducer of hepatic enzymes, reducing the effect of warfarin and oral contraceptives.
- *Resistance.* Resistant mutants emerge when used as a single agent; resistance is due to the change in a single amino acid of the RNA polymerase target site. Often used in combination with other agents to prevent the development of resistance.
- *Clinical applications.* Tuberculosis (as part of triple therapy); in combination with other antibiotics for endocarditis and osteomyelitis; as prophylaxis for close contacts of meningococcal and *H. influenzae* meningitis.

Nitroimidazoles

METRONIDAZOLE

- *Mechanism of action.* Metabolized by nitroreductases to active intermediates which result in DNA breakages.
- *Pharmacokinetics.* Well absorbed orally or *per rectum* and distributed into

most tissues; metabolized in the liver and metabolites excreted in the urine.
- *Antimicrobial spectrum of activity*. It is active against *Bacteroides* species; other anaerobes; *Giardia lamblia*, *Trichomonas vaginalis* and other parasites.
- *Toxicity*. Nausea, metallic taste; rarely peripheral neuropathy.
- *Resistance*. Rare, but may occur due to reduced uptake or decreased nitroreductase activity.
- *Clinical applications*. Treatment of anaerobic infections, giardiasis, trichomoniasis and amoebiasis.

OTHER ANTIBIOTICS

Fucidin

- Well-absorbed; good penetration into bone and joints.
- Active against staphylococci.
- Acts on bacterial ribosome; resistance develops rapidly if used as monotherapy.
- Toxicity causes hepatic damage.
- Used in combination with flucloxacillin for serious staphylococcal infection (e.g. osteomyelitis, endocarditis).

Nitrofurantoin

- Well-absorbed; excreted in urine.
- Used in uncomplicated urinary tract infections.

Antimycobacterial agents

- Rifampicin (see above).
- Isoniazid; side-effects include hepatitis and peripheral neuropathy.
- Pyrazinamide; may cause hepatic damage.
- Ethambutol; may cause optic neuritis.

RESISTANCE TO ANTIMICROBIAL AGENTS

Mechanisms

There are four main mechanisms by which bacteria become resistant to antimicrobial agents:

- alteration of the target site to reduce or eliminate binding of the drug to the target;
- destruction/inactivation of the antibiotic;
- blockage of transport of the agent into the cell;
- metabolic bypass, providing the cell with a replacement for the metabolic step inhibited by the drug.

ALTERATION OF THE TARGET SITE

This type of resistance is typical of antibiotics that act on the ribosome (e.g. erythromycin, fusidic acid and rifampicin). Resistance occurs when a bacterial mutant arises from within a previously sensitive bacterial population and the presence of the antibiotic then allows the resistant mutant to be selected out.

The alteration of penicillin-binding proteins is a further example of bacteria altering the target site, resulting in the organism becoming resistant to β-lactam antibiotics (e.g. methicillin-resistant *Staphylococcus aureus* and penicillin-resistant pneumococci). Despite the considerable usage of β-lactam antibiotics, this type of resistance remains uncommon.

DESTRUCTION/INACTIVATION OF ANTIBIOTIC

β-lactamases

β-lactamases cause hydrolysis of the β-lactam ring to form an inactive product. β-lactamase production may be chromosomally or plasmid mediated.

Aminoglycoside-modifying enzymes

Production of aminoglycoside-modifying enzymes is normally plasmid mediated.

INTERFERENCE WITH DRUG TRANSPORT

Resistance to some antibiotics involves interference with the transport of these drugs into the bacterial cell, e.g. tetracyclines, aminoglycosides.

METABOLIC BYPASS

The antibiotic trimethoprim blocks folate metabolism by inhibiting the enzyme dihydrofolate reductase (see Fig. 26.2). Bacterial strains resistant to trimethoprim synthesize a trimethoprim-insensitive dihydrofolate reductase (encoded on a plasmid), allowing the organism to bypass the action of the antibiotic. Sulphonamide resistance is mediated by a similar mechanism.

The genetics of bacterial resistance

Resistance of bacteria to antibiotics is based on genetic changes, which allow the organism to avoid the action of the antimicrobial agent. Antibiotic-resistant genes may be passed between bacteria by a number of different vectors.
- Duplication of the bacterial chromosome takes place during cell division.
- Plasmids are extra-chromosomal pieces of DNA containing genes that confer on the bacteria a number of different properties. These genes are often non-essential, but allow the organism to survive under a variety of different conditions, giving the organism a survival advantage. Resistance to antibiotics is one of a number of properties which may be conferred on bacteria by plasmids.
- Transposons are small genetic elements consisting of individual or small groups of genes. These genetic elements have no ability to replicate, but can move from one replicating piece of DNA to another, for example, from the bacterial chromosome to a plasmid, and vice versa. Transposons play an important part in the evolution of bacterial resistance.
- Bacteriophages are viruses that infect bacteria and may act as a vector in carrying bacterial genes from one bacterial cell to another.

The existence of mechanisms for transferring genetic information (horizontal transmission) means organisms do not need to evolve antibiotic resistance mechanisms only by the process of mutation and selection (vertical transmission).

Epidemiology of antibiotic resistance

The principle underlying the emergence and spread of antibiotic resistance amongst bacteria is that the prevalence of resistance is directly proportional to the amount of antibiotic used. This is illustrated by the increased antibiotic resistance found in countries with unrestricted use of antibiotics, in hospitals compared to the community, and in intensive care units compared to general wards. Problem areas in antibiotic prescribing include:
- the use of antibiotics without prescription;
- the uncontrolled use of antibiotics in agriculture;

- poor prescribing habits;
- the absence of antibiotic policies, particularly in hospitals.

PRINCIPLES OF ANTIMICROBIAL THERAPY

- The need for antibiotic therapy should be carefully considered; mild, self-limiting infections often do not require treatment.
- The choice of antimicrobial agent is dependent on a number of factors relating to the patient, the organism and the site of infection (Table 26.2).
- The route of administration, dose and length of treatment depend on the severity and type of infection. Intravenous antimicrobials are used for severe infections.
- Antibiotic combinations.

   (a) Antibiotics used in combination may be *additive* (combined effect equal to the sum of the individual agents), *synergistic* (combined effect greater than achieved with addition; e.g. sulphonamide plus trimethoprim, gentamicin plus penicillin), or *antagonistic* (drugs inhibit the action of each other).

   (b) The use of two (or more) antibiotics may broaden the spectrum of antimicrobial cover, e.g. cefuroxime and metronidazole may be used to treat abdominal sepsis to cover coliform organisms and anaerobes respectively.

   (c) Occasionally, antimicrobial agents are used in combination to prevent treatment failure and the development of resistance, e.g. triple therapy for tuberculosis.

FACTORS AFFECTING CHOICE

- Pharmacology
- Interaction with other drugs
- Penetration to site of infection
- Cost
- Toxicity
- Likely pathogens
- Antimicrobial spectrum of activity
- Age of patient
- Tolerance
- Underlying conditions (e.g. renal or hepatic failure)
- Patient allergies
- Pregnancy

Table 26.2 Factors affecting choice of antimicrobial agent.

(d) Some antibiotics contain a fixed combination of two agents, e.g. co-amoxiclav contains amoxicillin plus a β-lactamase inhibitor in the same preparation. The β-lactamase inhibitor protects amoxycillin from degradation by certain β-lactamase-producing bacteria.

- Antibiotic prophylaxis. Antibiotics can be used for prophylaxis to prevent infections. Prophylactic antibiotics are used most frequently for preventing infection following certain types of surgery (e.g. abdominal); antibiotic prophylaxis for other conditions is discussed in Chapter 46.

ADVERSE REACTIONS AND ANTIBIOTICS
(Table 26.3)

Antibiotics may be associated with side-effects common to all drugs, including allergic reactions and toxicity to various organs (e.g. hepatotoxicity, nephrotoxicity).

Specific complications associated with antibiotics relate to their depressive effect on normal commensal flora, e.g. oral candidiasis and *Clostridium difficile* colitis.

ADVERSE REACTIONS

Antimicrobial	Adverse reaction
Penicillins	Allergic reactions
Aminoglycosides	Nephrotoxicity Ototoxicity
Vancomycin	Nephrotoxicity Ototoxicity
Sulphonamides	Folate deficiency Marrow depression Stephens–Johnson syndrome
Rifampicin Isoniazid Fucidin	Hepatotoxicity
Chloramphenicol	Aplastic anaemia

Table 26.3 Adverse reactions of some antimicrobials.

LABORATORY ASPECTS OF ANTIBIOTIC THERAPY

Assessment of bacterial sensitivity

A number of techniques are available to assess susceptibility of bacteria to antibiotics *in vitro*.

- *Disc diffusion tests.* Antibiotic-containing filter paper discs are placed on agar plates inoculated with the bacteria to be tested. After overnight incubation, susceptible bacteria will show a zone of growth inhibition around the disc; resistant bacteria grow up to the disc edge (Plate 45).
- *Minimum inhibitory concentration (MIC) assay.* These give a more detailed assessment of bacterial susceptibility. Suspensions of the test bacteria are incubated overnight with doubling dilutions of the antibiotic; the lowest concentration that inhibits growth is the MIC.
- *Minimum bactericidal concentration (MBC) assay.* This is similar to MIC but the end-point is bacterial killing rather than growth inhibition.

Therapeutic drug monitoring

This is required in patients receiving antimicrobials with a therapeutic threshold close to the toxic range; examples include aminoglycosides (e.g. gentamicin, tobramycin, amikacin) and glycopeptides (e.g. vancomycin). Results of serum monitoring are used to adjust subsequent dosing of the patient. Patients with renal or hepatic insufficiency may require monitoring of other antimicrobials.

CHAPTER 27

Antifungal Agents

Selective action against fungi, without toxic side-effects, is difficult to achieve and there is a limited number of useful antifungal agents. The principal ones used in clinical practice are the azoles, flucytosine, griseofulvin, and amphotericin B.

Azoles

- *Mechanism of action.* Azoles alter the fungal cell membrane by blocking biosynthesis of ergosterol resulting in leakage of cell contents.
- *Toxicity.* They may cause transient abnormalities of liver function; severe hepatotoxicity is a rare complication of ketoconazole therapy.

FLUCONAZOLE

- *Antifungal activity.* Active against *Candida* species (some species are resistant), *Cryptococcus neoformans* and *Histoplasma capsulatum*; ineffective against *Aspergillus* and *Mucor*.
- *Pharmacokinetics.* Oral and intravenous preparations are available; good penetration into various body sites. Fluconazole is excreted unchanged in urine.
- *Clinical applications.* Mucocutaneous and invasive candidiasis; antifungal prophylaxis in immunocompromised patients (e.g. AIDS and transplant patients); cryptococcal infections.

ITRACONAZOLE

- *Antifungal activity.* Active against *Candida, Cryptococcus neoformans, Histoplasma capsulatum, Aspergillus* and dermatophytes.
- *Pharmacokinetics.* Only available orally but well absorbed. It is highly protein-bound and degraded into a large number of inactive metabolites and excreted in bile.
- *Clinical applications.* Dermatophytoses; mucocutaneous and invasive candidiasis; cryptococcal infections; aspergillosis; histoplasmosis; blastomycosis.

KETOCONAZOLE

Available as oral or topical forms and used principally as topical therapy for dermatophyte infections and cutaneous candidiasis.

MICONAZOLE

Available as topical, oral, and parenteral forms; is used principally as topical therapy for dermatophyte infections and cutaneous candidiasis.

Flucytosine (5-fluorocytosine)

- Synthetic fluorinated pyrimidine.
- *Mechanism of action.* Deaminated in fungal cells to 5-fluorouracil which is incorporated into RNA in place of uracil, resulting in abnormal protein synthesis and inhibition of DNA synthesis. Resistance may arise during treatment.
- *Pharmacokinetics.* Intravenous and oral forms available; good tissue and CSF penetration; excreted via the urine.
- *Antimicrobial spectrum of activity.* Effective against yeasts, including *Candida* and *Cryptococcus neoformans*.
- *Toxicity.* Marrow aplasia.
- *Clinical applications.* Combined with amphotericin B in the treatment of serious cryptococcosis and systemic candidiasis.

Griseofulvin

- *Mechanism of action.* Inhibition of nucleic acid synthesis and damage to cell wall by inhibiting chitin synthesis; resistance is rare.
- *Pharmacokinetics.* Available in oral preparation only; well absorbed and concentrates in keratin, therefore useful in the treatment of dermatophytosis. Griseofulvin is metabolized in the liver and metabolites are excreted in urine.
- *Antimicrobial spectrum of activity.* Restricted to dermatophytes.
- *Toxicity and side-effects.* Generally well tolerated; urticarial rashes may occur.
- *Clinical applications.* Dermatophytoses.

Amphotericin B

- *Mechanism of action.* Damages the fungal cell membrane by binding to sterols; this results in leakage of cellular components and cell death.

- *Pharmacokinetics.*
  (a) Topical preparations (lozenges/mouth washes) for oral candidiasis.
  (b) Parenteral preparation for intravenous administration; highly protein-bound with low penetration into many body sites, including cerebrospinal fluid (CSF). Metabolized in the liver.
- *Antimicrobial activity.* Broad spectrum of activity, including *Aspergillus, Candida, Blastomyces, Coccidioides, Cryptococcus neoformans, Histoplasma capsulatum* and *Mucor*. Resistance is rare.
- *Toxicity.*
  (a) Anaphylactic reactions occur in some patients (a small test dose should be given before commencing therapeutic doses); fever and chills common.
  (b) Renal tubular damage is a serious problem; renal function should be monitored regularly; doses should be increased slowly.
  (c) Recently, liposomal amphotericin preparations (amphotericin complexed to microscopic lipid micelles) have been developed which allow higher doses of amphotericin to be used without toxicity.
- *Clinical applications.* Amphotericin remains the most effective therapy for systemic mycoses, including disseminated candidosis, cryptococcosis, aspergillosis and mucormycoses.

CHAPTER 28

Antiviral Agents

Several antiviral agents are currently available (Table 28.1) and many new drugs are being developed, very probably resulting in an increase in therapy for viral infections.

MECHANISMS OF ACTION

There are three broad groups:
1 those that directly inactivate intact viruses (viricidal) include ether, detergents, ultraviolet light; these are not useful for treating patients (see Disinfectants);
2 those that inhibit viral replication at cellular level (antivirals);
3 those that modify host responses (immunomodulators).

Antiviral agents inhibit:
- virus uncoating;
- viral genome transcription and replication;
- virus maturation;
- virion assembly.

Immunomodulators may enhance deficient host-immune respone.

DRUG RESISTANCE

Drug resistance results from point mutations within the viral genome. Most drug-resistant viruses are found in immunocompromised patients and include HIV, varicella-zoster virus, herpes simplex virus, and cytomegalovirus.

The main clinically valuable antiviral agents are presented below.

Amantadine/rimantadine

- *Structure.* Tricyclic amines.
- *Mechanism of action.* They inhibit an early step (uncoating) in the replication of influenza A virus.
- *Toxicity and side-effects.* Gastrointestinal upset, lack of concentration; in

AVAILABLE ANTIVIRAL AGENTS

Group	Drug	Disease
Adamantanes	Amantadine	Influenza A
	Rimantadine	Influenza A
Phosphonates	Foscarnet	CMV retinitis
Interferons	Interferon-alpha	Chronic hepatitis C
		Chronic hepatitis B
Nucleoside analogues	Vidarabine	CMV—conjunctivitis
	Acyclovir	Herpes simplex virus infection, including genital herpes, encephalitis, skin lesions, varicella-zoster virus infection
	Ganciclovir	CMV—pneumonia
	Zidovudine	HIV disease
	Ribavirin	RSV disease

CMV, cytomegalovirus; HIV, human immunodeficiency virus; RSV, respiratory syncytial virus.

Table 28.1 Available antiviral agents for treating various infections.

renal failure or at a high dose amantadine can cause seizures, coma, cardiac arrhythmias.
- *Resistance.* Readily selected; up to 30% of patients have a resistant virus following treatment.
- *Clinical applications.* Mainly prevention of influenza A virus infection in certain groups of patients, including the immunocompromised, who cannot tolerate vaccine or mount an adequate immune response.

Foscarnet

- *Structure.* Foscarnet is an inorganic pyrophosphate analogue. It inhibits herpes viruses and most ganciclovir-resistant cytomegalovirus and acyclovir-resistant herpes simplex virus and varicella-zoster virus mutants.
- *Mechanism of action.* Foscarnet directly inhibits herpes virus DNA polymerases.
- *Toxicity and side-effects.* Nephrotoxicity; hypo- and hypercalcaemia, hypokalaemia. Central nervous system effects range from headache and seizures (up to 10%), to nausea and anaemia.
- *Resistance.* Herpes simplex virus, varicella-zoster virus and cytomegalovirus can develop resistance.

- *Clinical applications.* Cytomegalovirus infections, acyclovir-resistant herpes simplex and varicella-zoster virus infections.

Interferons

Interferons are host cell proteins synthesized by eukaryotic cells which can inhibit growth of viruses. There are three major classes: α, β, and γ.
- *Mechanism of action.* Interferons have complex direct antiviral effects and enhance the immune response to viral infection.
- *Toxicity and side-effects.* Influenza-like illness; bone marrow suppression; confusion, coma, liver and renal damage.
- *Clinical applications.* Chronic hepatitis B and C; other possible applications under study include herpes virus, HIV, papillomavirus infections.

Vidarabine

- *Structure.* Vidarabine is an analogue of adenosine with activity against herpes (herpes simplex virus and varicella-zoster virus). It has less activity against CMV and Epstein–Barr virus. Vidarabine inhibits viral DNA synthesis.
- *Toxicity and side-effects.* Gastrointestinal and neurologic.
- *Resistance.* Resistant HSV mutants have been isolated.
- *Clinical applications.* Herpes simplex virus encephalitis, neonatal herpes; herpes simplex virus keratitis; zoster or varicella in the immunocompromised.

Acyclovir has largely replaced vidarabine due to greater safety and efficacy.

Acyclovir

- *Structure.* A deoxyguanosine analogue, acyclovir is active against herpes simplex and varicella-zoster viruses.
- *Mechanism of action.* It is activated by viral thymidine phosphokinases and cellular kinases. It inhibits viral DNA polymerases and blocks viral DNA synthesis.
- *Toxicity and side-effects.* Cellulitis at injection site; rash; haematuria, nausea, neurotoxicity; confusion, tremors, hallucinations, seizures ($<4\%$); reversible renal dysfunction ($<5\%$). But in general high chemotherapeutic index.
- *Resistance.* This occurs with herpes simplex virus and varicella-zoster viruses. There are several mechanisms, including lack of viral thymidine kinase and altered DNA polymerase.

- *Clinical applications.* Herpes simplex virus keratitis, genital herpes, severe varicella-zoster infection; HSV encephalitis.

Famciclovir

- *Structure.* Famciclovir is an acyclic nucleoside analogue.
- *Mechanism of action.* This is similar to acyclovir. It inhibits viral DNA synthesis and is active against herpes viruses.
- *Toxicity and side-effects.* Headache, nausea.
- *Resistance.* This can occur, but is ill-defined at present.
- *Clinical applications.* Treatment of severe herpes zoster and genital herpes simplex.

Ganciclovir

- *Structure.* Ganciclovir is a deoxyguanosine analogue.
- *Mechanism of action.* Following intracellular phosphorylation it inhibits viral DNA polymerases.
- *Toxicity and side-effects.* Myelosuppression with: neutropaenia (up to 40% of patients); thrombocytopaenia (up to 20% of patients); effects are reversible on cessation; need to monitor blood counts. Central nervous system effects range from headache to coma (rare).
- *Resistance.* Occurs.
- *Clinical applications.* CMV infections, particularly in immunocompromised patients; prevention of CMV disease following organ transplantation.

Zidovudine

- *Structure.* A thymidine analogue, zidovudine (as named AZT, azidothymidine) inhibits various retroviruses, including HIV-1 and HIV-2.
- *Mechanism of action.* It inhibits viral RNA-dependent DNA polymerase (reverse transcriptase).
- *Toxicity and side-effects.* Granulocytopaenia, anaemia; central nervous system effects; headache, myalgia, seizures.
- *Resistance.* This develops gradually, resulting in high-level resistant HIV mutants in AIDS patients
- *Clinical applications.* HIV infection and post-exposure prophylaxis; the value of the latter is under debate.

Ribavirin

- *Structure.* A guanosine analogue, ribavirin inhibits a wide range of viruses, including herpesviruses and adenoviruses.
- *Mechanism of action.* This is not well defined. Ribavirin interferes with viral messenger RNA production.
- *Toxicity and side-effects.* Anaemia due to haemolysis or bone marrow suppression.
- *Resistance.* Not recognized.
- *Clinical applications.* RSV, bronchiolitis, pneumonia.

Others

- Dideoxycytidine (ddC), dideoxyadenosine (ddA); nucleoside analogues; used in combination with AZT in HIV disease.
- Proteinase inhibitors; inhibit HIV-specific proteinase; used in combination with other HIV-specific antivirals in HIV disease.

CHAPTER 29

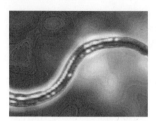

Skin and Soft Tissue Infections

NORMAL FLORA OF SKIN

The normal skin is colonized mainly by Gram-positive organisms (e.g. *Staphylococcus epidermidis*), which can survive the relatively harsh conditions: dryness, acidity and the presence of fatty acids and salt. The normal flora is protective, as other species must compete to survive in this ecological niche.

PATHOGENESIS OF SKIN AND SOFT TISSUE INFECTIONS

- Superficial skin infections affect the epidermis and dermis (e.g. impetigo, erysipelas).
- Deeper infections affect the subcutaneous tissue (e.g. cellulitis, fasciitis).
- Infections of skin structures may occur (e.g. hair follicles, folliculitis).
- Skin infections may arise from pathogens that gain entry through damaged areas of skin or from haematogenous spread.
- Bacterial toxins produced by infections at other sites may also result in skin changes (e.g. Group A streptococci causing scarlet fever).

SKIN AND SOFT TISSUE INFECTIONS

Folliculitis, furuncles and carbuncles

Aetiology. These are commonly caused by *S. aureus*; perianal lesions are often caused by a mixture of faecal organisms (e.g. *Escherichia coli*, anaerobes, *Streptococcus milleri*).

Pathogenesis. Folliculitis results from infection of a hair follicle (Plate 49). The lesion may expand and develop into a furuncle (boil). Boils may eventually rupture, with drainage of pus, or develop into carbuncles (large loculated abscesses), particularly in diabetics. Perianal abscesses are a particular problem in neutropaenic patients.

Complications
- Local spread to bones, joints and other deep structures.
- Haematogenous spread, resulting in abscesses in various organs.

Laboratory diagnosis. Culture of aspirated pus.

Treatment. Drainage and antibiotics (e.g. flucloxacillin for *S. aureus*; metronidazole for anaerobes).

Impetigo

Impetigo is a superficial infection of the skin, involving the epidermis.

Aetiology. Group A β-haemolytic streptococcus frequently in association with *S. aureus*.

Epidemiology. Principally associated with childhood, poor socioeconomic conditions and overcrowding.

Clinical features. Yellow crusted lesions develop, mainly on the face. The lesions spread by auto-inoculation.

Laboratory diagnosis. Culture of swabs of lesions.

Treatment. Penicillin or flucloxacillin are the drugs of choice.

Acne

Acne is associated with *Propionibacterium acnes*. Lesions occur in sebaceous follicles with plugs of keratin blocking the sebaceous canal, resulting in 'blackheads'. Treatment is by tetracycline combined with topical antiseptics.

Erysipelas

Erysipelas is a streptococcal infection involving the superficial layers of the skin (epidermis and dermis).

Epidemiology. Erysipelas principally occurs in elderly patients.

Pathogenesis. Group A β-haemolytic streptococci gain entry through breaks in the skin.

Clinical features. Erysipelas most commonly affects the face, particularly the cheeks and peri-orbital areas (butterfly rash) (Plate 50). The area of erythema shows a clear line of demarcation, which helps to distinguish erysipelas from cellulitis. Fever is common.

Laboratory diagnosis. Isolation of the Group A streptococcus from superficial swabs is often unsuccessful; blood cultures and skin aspirates may be positive in about one-quarter of patients. Serological evidence (e.g. anti-streptolysin O (ASO) titre) of streptococcal infection may be helpful.

Treatment. Benzylpenicillin or erythromycin.

Cellulitis

Cellulitis is an infection of the dermis and subcutaneous fat.

Aetiology
- Group A β-haemolytic streptococcus is the most common cause of cellulitis (Plate 51). Rare complications include septicaemia and acute glomerulonephritis. Other β-haemolytic streptococci may also cause cellulitis.
- *Haemophilus influenzae* type b is a cause in children <5 years.
- *Erysipelothrix rhusiopathiae* is an unusual cause of cellulitis, found in meat and fish handlers.
- Non-cholera vibrios (e.g. *Cholera vulnificus*) are a rare cause of cellulitis which complicates wounds that have become contaminated with seawater.
- Anaerobic cellulitis is a complication of bites (oral flora anaerobes) or infection of devitalized tissue (clostridia) following trauma, diabetes mellitus, or vascular insufficiency (Plate 52).

Pathogenesis. Bacteria gain entry through a break in the skin, although this may not be evident at the time of presentation.

Clinical features. Erythematous, hot, swollen skin is seen with an irregular edge (contrast with erysipelas). There is fever and enlargement of regional lymph nodes. Anaerobic cellulitis may be foul smelling, with skin crepitus and the presence of necrotic skin.

Laboratory diagnosis. Culture of skin swabs/aspirates (positive in approximately one-quarter of cases) and by blood cultures (occasionally positive).

Treatment
- For streptococci/*Erysipelothrix*, penicillin is given. (*Note*: staphylococcal superinfection of cellulitis is not uncommon and requires treatment with flucloxacillin.)
- Anaerobes are treated with metronidazole and debridement of devitalized tissue.
- Treatment of *H. influenzae* type b is with cefotaxime.

Synergistic bacterial gangrene

This is an infection of the superficial fasciae and subcutaneous fat, not involving muscles. It may involve the abdominal wall or male genitalia (Fournier's gangrene).

Aetiology/pathogenesis. Frequently mixed infection occurs involving *S. aureus*, microaerophilic or anaerobic streptococci, and, occasionally, Gram-negative anaerobes (e.g. *Bacteroides*). Predisposing conditions include local trauma, abdominal and genital surgery, and diabetes mellitus.

Clinical features. Rapid, spreading cellulitis is present with black necrotic areas. The condition is painful, with a high fever.

Laboratory diagnosis. Culture of affected tissue or pus.

Treatment. Radical excision of the necrotic area is required plus antibiotic therapy: flucloxacillin (*S. aureus*); penicillin (streptococci); metronidazole (anaerobes).

Necrotizing fasciitis

An infection of the skin and deeper fascia; may involve muscle.

Aetiology. Group A β-haemolytic streptococci, occasionally with anaerobes. Onset may follow recent surgery.

Pathogenesis. Group A streptococci produce enzymes which facilitate tissue spread.

Clinical features. Rapidly spreading cellulitis with necrosis; the patient is toxic with elevated temperature; high mortality.

Laboratory diagnosis. Culture of affected tissue or pus.

Treatment. Resection of affected tissue plus high-dose benzylpenicillin with metronidazole; clindamycin is an alternative in penicillin-allergic patients. Resection is also necessary.

Clostridial cellulitis and clostridial myonecrosis (gas gangrene)

Epidemiology. Now rare in developed countries; peripheral vascular disease and diabetes mellitus are important risk factors.

Pathogenesis. Wounds following trauma or surgery can become infected with clostridia, in particular, *Clostridium perfringens*, resulting in gas gangrene or clostridial myonecrosis.

Clostridia produce enzymes and exotoxins which facilitate rapid spread through tissue plains and result in haemolysis and shock; gas production results in tissue crepitus.

Clinical features. Cellulitis with necrotic areas and gangrene, characterized by tissue crepitus, fever and toxaemia.

Laboratory diagnosis. Isolation of clostridia from tissue/swabs or blood. (*Note*: the isolation of clostridial species from wounds does not define a case of gas gangrene; clostridia can contaminate wounds without infection; gas gangrene is primarily a clinical diagnosis.)

Treatment. Penicillin or metronidazole plus debridement of devitalized tissue.

Prevention. Wounds with devitalized tissue should be debrided, cleaned thoroughly and carefully monitored during healing. Prophylactic antibiotics (penicillin) should be prescribed for patients undergoing amputations for peripheral vascular disease.

Venous ulcers/bedsores

A variety of conditions (e.g. diabetes mellitus, peripheral vascular disease, venous insufficiency) may result in areas of skin which have a reduced blood supply and become necrotic, forming ulcers. Ulcers may occur around the foot and ankle (venous ulcers) or the sacral area in bedridden patients (bedsores) and are prone to bacterial colonization and, occasionally, infection.

Typical culture results include:
- mixed growth of coliforms and anaerobes; normally not significant;
- *P. aeruginosa* species; may reflect antibiotic replacement flora; occasionally requires antibiotic therapy;
- *S. aureus*; can be colonization or infection;
- β-haemolytic streptococci Groups C and G β-haemolytic streptococci may colonize venous ulcers and occasionally cause infection; Group A β-haemolytic streptococci frequently cause infection.

Distinguishing colonization from infection is important, as antibiotic treatment of colonized ulcers is of no benefit; spreading cellulitis around the ulcer suggests infection.

Burns

As with venous ulcers, burns may become colonized with bacteria, e.g. *S. aureus* and *Pseudomonas* species. The need for antibiotic treatment is dependent on the patient's condition, e.g. the viability of skin grafts, and the organism isolated (Group A β-haemolytic streptococci always require treatment).

Wound infections

Aetiology
- Accidental wounds: *S. aureus*, β-haemolytic streptococci, *Clostridia*.
- Bites: oral commensal bacteria, including anaerobes and streptococci; *Pasteurella multocida* (animal bites).
- Surgical wounds (see below).

Laboratory diagnosis. Culture of pus or wound swab.

Treatment. This is dependent on culture results. Prompt treatment of human and animal bites is important as organisms may be inoculated into deep tissues, with rapid spread of infection through tissue.

Surgical wound infections

Predisposing factors
- *Underlying conditions*: e.g. diabetes mellitus and peripheral vascular disease, which result in poor blood supply.

- *Type of operation*: surgery involving contaminated areas (e.g. gastrointestinal tract) or infected tissue (e.g. appendicitis; Plate 51) is more likely to result in post-operative wound infections.
- *Surgical technique*: e.g. poor technique in wound closure may result in haematoma formation or devitalized tissue, predisposing to wound infection.
- *Post-operative factors*: cross-infection from other infected cases.

Aetiology
- Upper body: S. aureus, Group A β-haemolytic streptococci.
- Lower body (particularly abdominal surgery): coliforms, occasionally mixed with anaerobes; P. aeruginosa.

Clinical features. Fever; wound exudate; cellulitis.

Laboratory diagnosis. Culture of wound swab or pus.

Treatment. Dependent on the causative organism.

Prevention
- *Pre-operative preparation*: treatment of sepsis before operation when possible; decontamination of operative area (e.g. bowel preparation) and the use of *prophylactic antibiotics* for operations involving contaminated or infected areas.
- *Good surgical technique* to prevent haematoma formation and wound necrosis and *aseptic theatre practice* with sterile instruments and a clean environment.
- *Measures to prevent post-operative contamination of wounds*, e.g. 'no-touch' dressing technique.

Toxin-mediated skin conditions

SCARLET FEVER

This is a Group A streptococcal throat infection caused by a strain producing an erythrogenic toxin which results in an erythematous rash with a red ('strawberry') tongue and circumoral pallor. As the rash fades there is desquamation of the skin. Treatment is penicillin or erythromycin.

STAPHYLOCOCCAL SCALDED SKIN SYNDROME

This occurs in small infants ('Ritter's diseases') and occasionally in older children and adults ('toxic epidermal necrolysis' or 'Lyell's disease'). The

S. aureus strain causes an infection at a distant site but produces a toxin which results in a severe rash with skin desquamation. Diagnosis is on clinical grounds and is supported by the isolation of a toxin-producing strain of *S. aureus*. Treatment is with flucloxacillin. A similar toxin-mediated rash may be seen in *toxic shock syndrome*, also a complication of *S. aureus* infection.

Dermatophyte fungal infections

These have a worldwide distribution, with transmission by direct contact or via fomites (e.g. towels, hairbrushes). Infection may be acquired from animals. See Table 29.1 for aetiology and clinical features.

Laboratory diagnosis. Microscopy and culture of appropriate specimens is made (skin scrapings; nail clippings; hair roots).

Treatment. This is with topical azole antifungals oral griseofulvin.

COMMON DERMATOPHYTE INFECTIONS

Infection	Site	Dermatophyte	Clinical features
Tinea corporis	Trunk Limbs	*E. floccosum* *Trichophyton* spp. *Microsporum* spp.	Pruritus; circular erythematous lesion, scaling, clearing from centre
Tinea cruris	Groin	*E. floccosum* *Trichophyton rubrum*	Pruritus; lesions similar to Tinea corporis
Tinea pedis	Feet	*E. floccosum* *Trichophyton* spp.	Pruritus; cracking and scaling between toes
Tinea unguium	Nails	*Trichophyton* spp.	Thick, yellowish nails with surrounding erythema
Tinea capitis	Hair	*Trichophyton tonsurans* *Microsporum audounii* *Microsporum canis*	Erythematous patches on scalp with scaling; hairs break, leaving bald patches

Table 29.1 Common dermatophyte (ringworm) infections.

Candidiasis

This affects moist skin areas (groin, axilla, skin folds, nailfolds) or damaged skin (e.g. irritant dermatitis in nappy area of babies). It causes erythema with vesicles and pustules. Treatment is with topical or oral azole antifungals (e.g. miconazole, fluconazole).

Pityriasis versicolor

This is a mild skin infection caused by the dimorphic fungus, *Malassezia furfur*. It is found worldwide, but is most common in the tropics. Dark, scaling lesions appear, particularly on the upper body. In dark-skinned patients, the affected areas become depigmented. Treatment is with topical or oral azole antifungals.

INFECTIONS OF THE EYE

EYELID

Eyelash follicle infections (styes) are usually caused by *S. aureus*.

ORBITAL CELLULITIS

Cellulitis of periorbital tissues caused principally by β-haemolytic streptococci and *H. influenzae* type b in children (<5 years) (Plate 50).

CONJUNCTIVITIS

- *Viral conjunctivitis*: aetiology includes adenoviruses and measles virus. Laboratory diagnosis is by viral culture. There is no specific treatment.
- *Neonatal conjunctivitis* (ophthalmia neonatorium) is described in Chapter 41.
- *Bacterial causes* include *S. aureus*, *H. influenzae* and pneumococci. It is most common in young infants. Treatment with topical antibiotics is normally effective.

HERPESVIRUS INFECTIONS OF THE EYE

- *Herpes simplex virus*: conjunctivitis and painful corneal ('dendritic') ulcers occur; these may result in corneal scarring. Latency with recurrent infection may occur. Treatment is with acyclovir or idoxuridine eyedrops.
- *Varicella-zoster virus*: ophthalmic zoster; the virus is latent in ophthalmic ganglia; it reactivates to affect the peri-orbital area (unilateral). Treatment is with acyclovir.
- *Cytomegalovirus*: choroidoretinitis; associated with AIDS. Treatment is with ganciclovir.

TRACHOMA

Chlamydia trachomatis infection of the eye (conjunctivitis and corneal lesions). It is prevalent in developing countries in the tropics, where it is an important cause of blindness. Spread is by direct hand–eye contact. Treatment is with tetracycline.

OTHER EYE INFECTIONS

- *Pseudomonas aeruginosa* may cause serious deep eye infections following trauma or surgery. Treatment includes parenteral antibiotics.
- *Toxoplasma gondii* causes choroidoretinitis, normally as part of congenital toxoplasmosis.
- *Toxocara canis* causes choroidoretinitis.
- *Onchocerciasis* is a nematode infection. Microfilariae invade the anterior chamber of the eye. It is an important cause of blindness in endemic areas (West Africa, Central America).
- *Acanthamoeba* is an amoebic infection causing keratitis. It is associated with contact lens use.

CHAPTER 30

Bone and Joint Infections

The aetiology, diagnosis and management of osteomyelitis (infection of bone) and septic arthritis are similar. Both conditions can present as acute or chronic infections.

OSTEOMYELITIS

Pathogenesis. Causative organisms may reach bone via haematogenous spread, directly from surrounding infected tissue, or from the environment following trauma (e.g. open fracture). Osteomyelitis following haematogenous spread is found most commonly in children and generally affects the growing ends of long bones. Osteomyelitis secondary to spread from an adjacent area of infection is more common in older patients (includes wound infections following orthopaedic surgery, mastoiditis), patients with peripheral vascular disease or diabetes.

Acute osteomyelitis

Aetiology. The aetiology of acute osteomyelitis is shown in Table 30.1.

Clinical presentation. There is high fever with raised white cell count and pain and tenderness over affected bone. In infants reduced use of the affected limb is an important sign. X-ray changes are often not apparent until 1–2 weeks after clinical presentation (Plate 53).

Laboratory diagnosis. Blood cultures are positive in approximately 50% of cases. Direct aspiration of bone is performed occasionally. Intra-operative specimens may be taken.

Management. Antibiotic treatment is normally for at least 4 weeks, but shorter courses are sometimes used (Table 30.1). Surgery to allow drainage and debridement of dead bone may be necessary.

ACUTE OSTEOMYELITIS

Organism	Treatment	Comment
Staphylococcus aureus	Flucloxacillin and fucidin	>90% of cases
Streptococcus pyogenes	Benzylpenicillin	Infrequent cause
Streptococcus pneumoniae	Benzylpenicillin	Usually direct spread from ear or sinus infection
Clostridium perfringens	Benzylpenicillin	Usually post-traumatic
Haemophilus influenzae type b	Cefotaxime	Children <5 years
Salmonella spp.	Ciprofloxacin	Rare cause in developed countries

Table 30.1 Common causative organisms of acute osteomyelitis.

CHRONIC OSTEOMYELITIS

Organism	Comment
Staphylococcus aureus	(>60% of cases)
Streptococcus milleri	Associated with abscesses
Streptococcus pneumoniae	Ear or sinus infection
Actinomyces israelii	Peridontal sepsis
Pseudomonas; coliforms	Contamination of open wounds
Salmonella	Associated with sickle-cell disease and septicaemia
Mycobacterium tuberculosis	Important cause
Bacteroides	Associated with chronic otitis media and mastoiditis

Table 30.2 Common causative organisms of chronic osteomyelitis.

Chronic osteomyelitis

Aetiology (Table 30.2). Often results from spread from a local adjacent infection, for example, chronic otitis media, mastoiditis or peridontal sepsis.

Clinical manifestations. Often these are related to the precipitating infection. Bone involvement may only become apparent following X-ray changes. Occasionally presents as chronic pain.

Laboratory diagnosis. As for acute osteomyelitis.

Management. Antibiotic treatment depends on the causative organisms. Surgical debridement of dead tissue is necessary more frequently than with acute osteomyelitis.

Tuberculous osteomyelitis

- Tuberculous osteomyelitis results following haematogenous dissemination of the organism during the early phases of infection.
- Presentation is insidious with no acute signs; pain is the most common symptom. A discharging sinus may be present occasionally. The most frequent sites are the weight-bearing bones, particularly the spine.
- Investigations include microscopy (Ziehl–Neelsen stain) and culture of fine-needle biopsy or surgical specimens.
- Treatment with antituberculous drugs is normally sufficient, although, occasionally, surgical intervention is required.

Osteomyelitis following trauma and surgery

Severe compound fractures may become infected with skin bacteria (e.g. staphylococci) or environmental organisms (e.g. clostridia). Contaminated compound fractures may require debridement of devitalized tissue and antibiotic therapy.

Osteomyelitis following orthopaedic operations is often caused by *S. aureus*, but, when associated with prosthetic joint replacements, is more frequently caused by *S. epidermidis* with a characteristic insidious onset and minimal symptoms. Antibiotic treatment is difficult as many *S. epidermidis* are often resistant to flucloxacillin. Prevention of orthopaedic prosthetic device infection involves the use of specially designed operating theatres ventilated with ultra-clean air, and perioperative prophylactic antimicrobials.

SEPTIC ARTHRITIS

Infection of joints may arise from haematogenous spread or following direct spread from an overlying skin lesion or area of osteomyelitis.

Aetiology
- *S. aureus*: 90% of cases.
- *H. influenzae* type b: the most common cause in children aged <5 years.
- Less common causes include β-haemolytic streptococci, salmonellae,

Neisseria gonorrhoea, *Streptococcus pneumoniae* and *Mycobacterium tuberculosis*.

Clinical presentation. There is a hot, swollen joint with fever (Plate 54). Tuberculous arthritis presents insidiously with a cold swollen joint.

Laboratory diagnosis. Blood cultures are positive in approximately 50% of cases. Joint aspiration is frequently performed for microscopy and culture.

Management. Antibiotic therapy is required (see Osteomyelitis) and surgical drainage may be necessary.

Reactive arthritis

This is immunologically mediated, resulting from infection at a distant site, e.g. *Yersinia* gastroenteritis, non-specific urethritis caused by *Chlamydia trachomatis*. Usually, more than one joint is affected.

Viral arthritis

A number of viruses may cause arthritis as part of a generalized infection, e.g. rubella virus, mumps virus and parvovirus.

CHAPTER 31

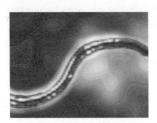

Oral and Dental Infections

NORMAL ORAL FLORA

The normal microbial flora (Table 31.1) of the oral cavity is influenced by many factors, including diet, age, hygiene and infection. Additional organisms found in saliva include *Nocardia* and *Actinomyces*. In comparison to saliva, fewer streptococci are associated with the surface of the teeth and gingival crevices.

DENTAL INFECTIONS

The normal flora of the oral cavity is responsible for a large proportion of oral infections (e.g. caries, abscesses, candidiasis). Lesions in the mouth can also be associated with generalized infections (e.g. Koplik spots in measles).

Plaque contains bacteria, inorganic material, including protein and polysaccharides, which adhere to teeth.

Dental caries

The role of bacteria in causing dental caries is supported by two theories:
• acidogenic theory: tooth enamel is damaged by acids, resulting from bacterial metabolism of carbohydrates;
• proteolytic theory: bacterial proteolytic enzymes directly attack enamel.

Organisms associated with caries include *Streptococcus mutans*, *S. sanguis* and some strains of lactobacilli and *Actinomyces*.

Prevention of caries includes: improving resistance of teeth (e.g. by the use of fluoride); modifying flora of plaque by the use of antimicrobials (e.g. chlorhexidine mouth rinses); and prevention of acid formation by removing fermentable substrates and sucrose from the diet.

ORGANISMS IN SALIVA		
Group	%	Examples
Gram-positive facultative cocci	50	Primarily viridans streptococci (e.g. *Streptococcus sanguis, S. salivarius*)
Gram-positive facultative bacilli	10	Diphtheroids
Gram-positive anaerobic cocci	10	Peptostreptococci
Gram-negative anaerobic cocci	15	Veillonella
Gram-negative anaerobic bacilli	5	Fusiforms, *Bacteroides, Prevotella*

Table 31.1 Organisms present in saliva.

Periodontal disease

Bacteria are associated with inflammatory disease of the periodontium and can cause chronic periodontal disease. There is a correlation between poor oral hygiene and periodontal disease. Chronic periodontal disease is slowly progressive, with the periodontium being affected leading to tooth loss; development of this disease involves several stages:
- acute exudative vasculitis, which is found in the gingival sulcus with polymorph involvement;
- connective tissue infiltration with plasma cells, immunoglobulin and complement. Loss of collagen with fibrosis and scarring results with destruction of the periodontal tissue and alveolar bone.

BACTERIAL ASSOCIATED CHRONIC PERIODONTAL DISEASE

Anaerobic Gram-negative bacteria, e.g. *Prevotella melaninogenicus* (Plate 22) and *Fusobacterium nuciliarsum* produce various enzymes (e.g. neuraminidase, proteinases and collagenases) and cause gingivitis leading to periodontitis. The point of attachment of the junctional epithelium to the tooth retracts, resulting in destruction of collagen and bone. This is slowly progressive.

Acute and necrotic ulcerative gingivitis (ANUG)

This is associated with poor oral hygiene and various immunodeficiency states and affects mainly the interdental gingival papillae, resulting in ulceration and destruction of gingival margin. It is caused by a mixture of anaerobic organisms (e.g. *Treponema vincenti*, *Fusobacterium fusiforms* and *Prevotella melaninogenicus*; Plate 22). Treatment is with metronidazole.

Gingivitis

Gingivitis is inflammation of the gingivae caused by bacteria and associated plaque.

Juvenile periodontitis

This is associated with plaque accumulation and is caused by anaerobic, Gram-negative bacilli (e.g. *Prevotella* species).

INFECTIONS OF THE ORAL MUCOSA

Oral syphilis

Primary syphilis may present with a chancre (painless ulcer), usually on the lips, but occasionally on the tongue or tonsils. Swabs of the ulcer cannot be relied upon for diagnosis as non-pathogenic spiral treponemas may be present in the normal flora. Diagnosis is confirmed serologically. In secondary syphilis ulcers may develop on skin and mucous membranes, including the lips, tongue and palate. Tertiary syphilis can develop many years later, and involve leukoplakia on the tongue.

Acute sialoadenitis

This is a painful infection of the salivary glands which is caused by viruses and occasionally *S. aureus* and β-haemolytic streptococci. The parotid gland is usually involved.

Candidiasis

Candidiasis presents as white plaques on the tongue and cheek mucosa. It may occur following the use of broad-spectrum antibiotics. Chronic atrophic candidiasis may develop, with erythema and oedema of the palatal surface, particularly with dentures ('denture stomatitis').

DENTAL ABSCESSES

The grouping of dental abscesses includes acute alveolar, apical, dentoalveolar, periodontal and submucosa abscesses. They often result from pre-existing oral infection or trauma. These abscesses are often associated primarily with anaerobic bacteria and facultative streptococci.

ACTINOMYCOSIS

Actinomyces israelii may cause infection of the mandible and associated soft tissue. Sinus formation may occur, with 'sulphur granule' discharge which contains clumps of the organism. Treatment includes drainage and penicillin.

HERPETIC STOMATITIS

Herpetic stomatitis is an acute gingivostomatitis caused by herpes simplex, usually type 1. A large number of vesicles cover the lips, gingivae and tongue. It is usually self-limiting, within 10–14 days. The virus may remain latent in trigeminal nerve ganglia and present later as secondary infection, 'cold sores', with vesicular lesions commonly on the lip margins. Treatment is with acyclovir, particularly in the early stages.

HERPANGINA

Herpangina is a coxsackie A virus infection with crops of tiny ulcers on the oral mucosa (particularly the soft palate). These coalesce and fade within a week.

CHAPTER 32

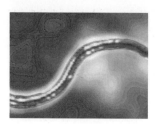

Upper Respiratory Tract Infections

Infections of the upper respiratory tract are common and are primarily of viral origin and rarely serious.

NORMAL FLORA

Many of the bacteria found as part of the normal flora of the mouth are also present in the pharynx. Other commensal bacteria include neisseriae, *Haemophilus* species and *Streptococcus pneumoniae*. *Staphylococcus aureus* is carried in the anterior nares of about 30% of humans and may also be isolated from the pharynx. β-haemolytic streptococci colonize the upper respiratory tract of about 5% of individuals.

VIRAL INFECTIONS

Common cold

Aetiology. The majority of colds are caused by rhinoviruses. Other associated viruses are influenza viruses, coronaviruses, coxsackie viruses, parainfluenza viruses, adenoviruses and echoviruses. A number of systemic viral infections may present with similar upper respiratory tract symptoms, e.g. measles, mumps and rubella.

Epidemiology. Transmission is by aerosol or via virus-contaminated hands. Nasal discharge causes irritation and results in sneezing, facilitating spread.

Clinical features. Nasal discharge often associated with cough and non-specific symptoms, such as headache and malaise; fever is not a prominent feature.

Laboratory diagnosis. Virus isolation and serological tests are usually not warranted because of expense, the benign course of the infection and the lack of an effective antiviral.

Treatment and prevention. There are no effective antiviral agents for treatment of the common cold viruses, except influenza; symptomatic treatment only. Influenza A and B vaccination can be given to aid prevention.

Viral pharyngitis, tonsillitis and glandular fever syndrome

Aetiology. Many of the viruses associated with the common cold syndrome may also result in pharyngitis including rhinoviruses, coronaviruses, coxsackie viruses, echoviruses, adenoviruses, influenza and parainfluenza viruses and respiratory syncytial virus (RSV). Epstein–Barr virus (EBV) and cytomegalovirus (CMV) cause glandular fever.

Epidemiology. Transmission is by aerosol and direct contact.

Clinical features
- Fever and pharyngeal discomfort with marked pharyngeal inflammation, occasionally with an exudate.
- The presence of vesicles on the soft palate indicates herpes simplex virus or coxsackie virus infection; conjunctivitis suggests influenza or adenovirus infection.
- Glandular fever syndrome: persistent pharyngitis with tiredness, malaise and cervical lymphadenopathy.

Laboratory diagnosis. Except in the case of glandular fever, laboratory diagnosis is rarely necessary. EBV and CMV infections may be diagnosed by specific serological assays.

Treatment. Symptomatic only.

Influenza

Aetiology. Influenza viruses A and B.

Epidemiology. Transmission is via aerosols or direct contact. Infection peaks during the winter months. Epidemics and pandemics of influenza are related to varying degrees of antigenic variation.

Clinical features. Nasal discharge, cough, conjunctivitis and fever are common. Complications of influenza are rare and include: primary viral pneumonia; secondary bacterial pneumonia (most frequently staphylococcal, now rare); encephalitis; Guillain–Barré syndrome; myocarditis and pericarditis.

Laboratory diagnosis. Virus isolation or serology is not attempted routinely, but is important in confirming influenza epidemics.

Treatment and prevention. Treatment is for symptomatic relief only; the antiviral agent amantadine is used occasionally in severe infections. Vaccines are available and are indicated for patients with: chronic cardiac and pulmonary disorders; patients with other chronic disorders (e.g. renal dysfunction, immunosuppression); and the elderly (>65 years).

Viral laryngotracheobronchiolitis (croup)

Aetiology. Parainfluenza viruses are the most frequent cause of croup; other viruses associated with upper respiratory tract infections can occasionally cause croup.

Epidemiology. Transmission is by aerosol spread. Most cases occur in the autumn and winter and the infection is restricted to children <5 years.

Clinical features. Distinctive deep cough ('bovine cough'), frequently with inspiratory stridor; dyspnoea and, in severe cases, cyanosis.

Laboratory diagnosis. Virus isolation (rarely performed).

Treatment. Symptomatic relief by mist therapy is often recommended but there is no scientific proof of its effectiveness; antibiotics are of no benefit except when croup is complicated by bacterial infection. In severe cases, hospitalization is required and ventilatory support may be necessary.

BACTERIAL INFECTIONS

Streptococcal pharyngitis

Aetiology. S. pyogenes and occasionally other β-haemolytic streptococci (Groups C and G).

Epidemiology. Approximately 5–10% of the population carry S. pyogenes in the pharynx; carriage rates are higher amongst children, particularly during the winter. Spread is by aerosol and direct contact, and is common within families; epidemics of streptococcal pharyngitis may occur occasionally in institutions, e.g. boarding schools.

Clinical features. Similar to viral pharyngitis; some features (e.g. systemic upset, purulent exudate) are claimed to be more prominent with streptococcal pharyngitis, but, in practice, it is difficult to separate viral from bacterial causes on clinical grounds alone.

Complications
- Local: peritonsillar abscess (quinsy), sinusitis, otitis media.
- Systemic: scarlet fever; rheumatic fever; acute glomerulonephritis, and septicaemia.

Laboratory diagnosis. Throat swabs are taken for culture. Direct antigen detection tests are available for rapid diagnosis. Serology (detection of antistreptolysin O antibodies) may be useful in patients presenting with post-streptococcal complications (rheumatic fever, glomerulonephritis).

Treatment. Penicillin, or erythromycin for penicillin-allergic patients.

Epiglottitis

Epiglottitis is an acute, severe infection of the epiglottis caused almost exclusively by *H. influenzae* type b.

Epidemiology. The infection is normally limited to children <5 years, and is spread by respiratory droplets.

Clinical features. There is acute onset of fever, sore throat and respiratory distress with stridor. Lateral X-rays of the neck may show the enlarged epiglottis. *Note*: when the disease is suspected, direct visualization of the epiglottis should *not* be attempted, except where facilities for immediate intubation are available.

Laboratory diagnosis. Isolation of the organism from throat swabs or blood cultures.

Treatment. Intubation and ventilation may be necessary. Intravenous antibiotic therapy should be instituted immediately; second or third generation cephalosporins (e.g. cefuroxime, cefotaxime) are the current treatment of choice. Ampicillin should not be used as sole therapy, as 10% of strains of *H. influenzae* produce β-lactamase.

Vincent's angina

This is an uncommon throat infection caused by a mixture of anaerobic bacteria (e.g. fusobacteria, spirochaetes) normally found in the mouth. As with acute ulcerative gingivitis, with which it may coexist, it is associated with poor oral hygiene and underlying conditions such as immunodeficiency.

Clinical features include pharyngitis with a necrotic pharyngeal exudate. Microscopy of the exudate shows typical fusobacteria and spirochaetes. Treatment is with penicillin or metronidazole.

Diphtheria

Aetiology. An upper respiratory tract infection, caused by toxin-producing strains of *Corynebacterium diphtheriae*, with central nervous system and cardiac complications mediated by the exotoxin.

Epidemiology. The organism is transmitted by aerosol spread. Following the introduction of vaccination, diphtheria is now rare in the developed world, but still common in developing countries.

Pathogenesis. The organism colonizes the pharynx, multiplies and produces toxin. The toxin acts locally to destroy epithelial cells and phagocytes resulting in the formation of a prominent exudate, sometimes termed a 'false membrane' (Plate 55). Cervical lymph nodes become grossly enlarged ('bull neck' appearance). The toxin spreads via lymphatics and blood, resulting in myocarditis and polyneuritis.

Clinical features. Pharyngitis with false membrane; enlarged cervical lymph nodes; myocarditis with cardiac failure; and polyneuritis.

Laboratory diagnosis. By isolation of the organism from throat swabs, followed by demonstration of toxin production by the Elek test.

Treatment and prevention. In suspected cases, treatment should be commenced before laboratory confirmation and includes antitoxin and antibiotics (penicillin or erythromycin). Close contacts should be investigated for the carriage of the organism, given prophylactic antibiotics (erythromycin) and immunized. Childhood immunization with diphtheria toxoid has resulted in the virtual disappearance of diphtheria from developed countries.

Acute otitis media

Aetiology and pathogenesis. Upper respiratory tract infection may result in oedema and blockage of the eustachian tube with subsequent impaired drainage of middle ear fluid predisposing to viral or bacterial infection (acute otitis media). About 50% are caused by respiratory viruses; common bacterial causes include *Strep. pneumoniae*, *H. influenzae*, β-haemolytic streptococci and *S. aureus*.

Epidemiology. It occurs worldwide and is most common in children <5 years with an increased incidence in winter months.

Clinical features. Fever and earache. The eardrum appears reddened and bulging and, if untreated, drum perforations with subsequent purulent discharge may occur.

Laboratory diagnosis. By culture of discharge. Needle aspiration of middle ear fluid (tympanocentesis) is performed occasionally.

Management. Antibiotics prescribed include amoxycillin or erythromycin. Follow-up is important as residual fluid in the middle ear ('glue ear') can result in hearing impairment.

Otitis externa

Infections of the external auditory canal are frequently caused by *S. aureus* and *Pseudomonas aeruginosa*. Topical treatment with antibiotic-containing eardrops is often effective.

Acute sinusitis

Acute sinusitis follows impaired drainage of the sinus cavity and may complicate viral upper respiratory tract infections. Bacterial causes are similar to otitis media. It presents with fever, facial pain and tenderness over affected sinuses; X-rays show fluid-filled sinuses. Management is with antibiotics (see Otitis media); occasionally sinus drainage is required.

CHAPTER 33

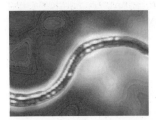

Lower Respiratory Tract Infections

PNEUMONIA

Pneumonia is relatively common, with a significant morbidity and mortality.

Definition. It is an acute respiratory infection with focal chest signs and radiographic changes (Plates 56–58). Various classifications are based on the causative organism, the radiographic appearance and histopathological changes. A practical classification is:
- community-acquired pneumonia;
- hospital-acquired pneumonia (nosocomial);
- pneumonia in the immunocompromised.

Community-acquired pneumonia

Epidemiology. This is an important infection worldwide. It is most common in the winter months; the overall incidence may vary in relation to outbreaks (e.g. *Legionella pneumophila*) or epidemics (e.g. *Mycoplasma pneumoniae*). Common viral causes of community-acquired pneumonia are shown in Table 33.1.

Clinical features
- *Symptoms and signs*: malaise, fever and shortness of breath; productive cough; pleuritic chest pain; tachypnoea, tachycardia; focal chest signs, e.g. consolidation; cyanosis.
- *Severe cases*: hypoxia, confusion, circulatory collapse resulting in renal and hepatic dysfunction.
- *Chest X-ray* (CXR): various appearances; lobar or patchy consolidation or diffuse shadowing.

Complications. Septicaemia and empyaema (infected pleural effusion).

Laboratory diagnosis
- Sputum specimens for microscopy (Gram stain) and culture.

COMMON VIRAL CAUSES OF PNEUMONIA	
Virus	Notes
Influenza viruses	Epidemics/pandemics; elderly and patients with chronic lung disease at risk
Parainfluenza viruses	Mainly in young children causing croup
Respiratory syncytial virus	Bronchiolitis in infants <6 months
Measles virus	Rare; seen mainly in adults and immunocompromised
Varicella-zoster virus	Rare; seen mainly in adults and immunocompromised
Adenoviruses	Young adults; associated with pharyngitis and conjunctivitis
Cytomegalovirus	Important in immunocompromised, particularly transplant patients
Herpes simplex virus	Immunocompromised and neonates

Table 33.1 Common viral causes of pneumonia.

- Bronchial-alveolar lavage specimens obtained by bronchoscopy for microscopy, culture and direct immunofluorescence tests.
- Blood cultures.
- Serology (viruses, *Mycoplasma*, *Chlamydia*, *Coxiella*, *Legionella*).

SPECIFIC CAUSES OF COMMUNITY-ACQUIRED PNEUMONIA
Streptococcus pneumoniae

This is a common cause of community-acquired pneumonia (30–50%) and occurs in all age groups. At-risk groups include patients with chronic lung disease, splenectomized patients and the immunocompromised. Vaccination is recommended for these groups. CXR commonly shows lobar consolidation. Laboratory diagnosis is by sputum and blood culture. Treatment includes penicillin or erythromycin.

Mycoplasma pneumoniae

Incidence varies (<1–20%) due to epidemics; occurs primarily in young adults. CXR commonly shows diffuse changes. Laboratory diagnosis is by serology. Treatment includes erythromycin.

Haemophilus influenzae

This accounts for <10% of all community-acquired cases. *H. influenzae* type b is an important cause of pneumonia in infants <5 years. Laboratory diagnosis is by sputum culture. Treatment is according to antibiotic susceptibility results.

Staphylococcus aureus

An important but now rare cause of post-influenza pneumonia the incidence is highest in small children. Laboratory diagnosis is by sputum and blood culture. Treatment is with flucloxacillin.

Legionella pneumophila

L. pneumophila is transmitted via aerosols particularly from contaminated air-conditioning systems. Less than 5% of pneumonias are caused by *L. pneumophila* but incidence varies according to outbreaks. Hospital outbreaks may occur; cases associated with recent travel can arise. The elderly and immunocompromised are particularly at risk. Laboratory diagnosis is by direct immunofluorescence on broncheolar-alveolar lavage specimens, sputum culture, serology.

Chlamydia pneumoniae

C. pneumoniae has only recently been recognized as a cause of pneumonia and the detailed epidemiology is still unclear. It probably causes about 10% of pneumonias with a higher incidence in young adults. Laboratory diagnosis is by serology.

Chlamydia psittaci

This is associated with contact with birds, including pigeons and parrots. Laboratory diagnosis is by serology.

Coxiella burnetii

C. burnetii is a rare cause of pneumonia and is usually confined to farm workers and vets. Laboratory diagnosis is by serology.

Viral pneumonia (Table 33.1)

Viral pneumonia occurs mainly in children, the elderly and immunocompromised patients, with an increased incidence in winter.

Hospital-acquired (nosocomial) pneumonia

This is a lower respiratory tract infection (fever, chest signs, radiological changes) presenting ⩾2 days after admission to hospital. Pneumonia is the third most common nosocomial infection, affecting about 0.5% of hospitalized patients. Risk factors include assisted respiration, immunocompromise and pre-existing pulmonary disease.

Aetiology. Commonly *Streptococcus pneumoniae*; *Haemophilus influenzae* are the cause, as well as Gram-negative organisms: *Escherichia coli*, *Klebsiella* spp., *Serratia* spp., *Pseudomonas aeruginosa*. These are anaerobes and are associated with aspiration of flora from the gastrointestinal tract causing aspiration pneumonia.

Pneumonia in the immunocompromised

Immunocompromised patients may become infected with classical chest pathogens, e.g. *S. pneumoniae*, *M. pneumoniae* or the important opportunistic lung pathogens (Table 33.2).

Laboratory diagnosis. This is often difficult; early bronchoscopy for bronchial-alveolar lavage fluid is important and should be examined by:
- microscopy (Gram and Ziehl–Neelsen stains);

PATHOGENS CAUSING CHEST INFECTIONS

Pathogen	Notes
Bacterial	
Actinomyceteae	Often cause abscesses and sinus formation
Atypical mycobacteria	Common in AIDS
Legionella pneumophila	May be hospital-acquired
Fungi	
Candida species	Often preceded by oral candidiasis
Aspergillus and *Mucor* species	Transmission by spores; outbreaks have occurred in association with hospital building work
Pneumocystis carinii	Common in AIDS
Viruses	
Cytomegalovirus	Important in transplant patients
Herpes simplex virus, varicella-zoster virus, measles virus	Important in immunocompromised children

Table 33.2 Pathogens causing opportunistic chest infections in immunocompromised patients.

- culture (bacteria, fungi, mycobacteria, viruses);
- direct immunofluorescence (*Pneumocystis carinii* and *Legionella*).

Serology: atypical pneumonias, including *M. pneumoniae* and *L. pneumophila*.

Treatment. In the absence of clear aetiology, empirical therapy is often required. Choice is dependent on clinical presentation, laboratory data, underlying disease and previous antibiotic therapy.

BRONCHIOLITIS

Bronchiolitis is caused by the respiratory syncytial virus (RSV) and results in obstruction of bronchioles by mucosal oedema. Infection occurs mainly in infants <18 months; it is particularly severe in infants <6 months. Spread is by aerosols. Clinical manifestations include fever, dyspnoea, respiratory distress, cyanosis. Laboratory diagnosis is by immunofluorescence for RSV antigen in nasopharyngeal aspirates or culture. Management may include nebulized ribavarin.

ACUTE EXACERBATION OF BRONCHITIS

This is principally viral (rhinoviruses, coronaviruses, influenza and parainfluenza viruses); rarely caused by *M. pneumoniae*. Clinical manifestations include dry cough, fever, malaise. Laboratory diagnosis is by culture or serology, but this is rarely performed. Treatment is with antibiotics if indicated.

CHRONIC BRONCHITIS

Pathogenesis. The condition results in increased susceptibility to infection by *H. influenzae*, *S. pneumoniae*, *Moraxella catarrhalis* and some viruses.

Clinical features. Chronic sputum production (often mucoid but it may be purulent) is part of the underlying condition. Acute-on-chronic infection occurs with increased production of purulent sputum and increasing dyspnoea.

Laboratory diagnosis. Sputum culture is performed for bacterial pathogens but results need careful interpretation.

Treatment. Antibiotic treatment should be prescribed in relation to the clinical picture and sputum culture results, and includes amoxycillin, tetracyline and co-amoxiclav.

BRONCHIECTASIS

There is underlying lung pathology with damage to the terminal bronchi, and bronchioles which act as a site of chronic infection. Pathogens include *H. influenzae, S. pneumoniae, Moraxella catarrhalis, P. aeruginosa* and anaerobes. Clinical features are similar to chronic bronchitis; antibiotic therapy is guided by sputum results and condition.

WHOOPING COUGH (PERTUSSIS)

Aetiology and epidemiology. Caused by *Bordetella pertussis*. Transmission is by aerosol. It is principally an infection of childhood but adults may occasionally be affected. Vaccination has dramatically reduced the incidence in many countries.

Clinical features. The incubation period is 7–21 days with a prodromal phase (7–14 days) with coryzal symptoms. A severe cough develops; bouts of coughing are frequently followed by an inspiratory whoop. Complications include chronic lung disease and cerebral damage due to violent coughing (rare).

Laboratory diagnosis. Culture of perinasal swabs is made.

Treatment and prevention. Treatment is with erythromycin, which reduces infectivity if given early, but has little effect on the clinical course. Live attenuated vaccine is given in childhood.

TUBERCULOSIS

Epidemiology. Tuberculosis remains a common infection in the developing world, particularly in Asia and Africa; the current AIDS epidemic has led to an increase in cases. The incidence of tuberculosis in Western Europe and North America has declined over the last 30 years due to improvements in nutrition, housing, and various preventive measures, including Bacille Calmette–Guérin (BCG) vaccination. More recently this decline has plateaued and in some developed countries a small increase in cases has been reported. In the UK, approximately 6000 new cases, with about 500 associated deaths, are recognized each year; infection is more common in certain groups, including Asian immigrants and in elderly men.

Pathogenesis. Primary tuberculosis: inhalation of *M. tuberculosis* results in a mild acute inflammatory reaction in the lung parenchyma, with phagocytosis of bacilli by alveolar macrophages. Bacilli survive and multiply within the macrophages and are carried to the hilar lymph nodes which enlarge. The local lesion and enlarged lymph nodes are called the primary complex (referred to as the 'Ghon focus').

Histologically, granulomas consisting of epithelial cells and giant cells develop, which eventually necrose (caseous necrosis). In many individuals, infection does not progress; however, organisms may remain viable for many years within lymph nodes. The local lesion and nodes become fibrotic and calcified.

In some patients, particularly the immunocompromised, organisms spread locally and via the blood stream to other organs, causing widespread disease (miliary tuberculosis).

Secondary tuberculosis: may arise in two ways:
- dormant mycobacteria may become reactivated, often due to lowered immunity in the patient; reactivation occurs most commonly in the lung apex (Plate 58), but may occur in other organs (e.g. kidney, bone);
- a patient may become reinfected following further exposure to an exogenous source.

As with primary tuberculosis, local and distant dissemination may occur.

Clinical features
- *Pulmonary tuberculosis*: chronic cough, haemoptysis, weight loss, malaise and night sweats. CXR: cavitating lesion, frequently in apical regions, with enlarged hilar lymph nodes.
- *Extrapulmonary tuberculosis*:
    (a) *genitourinary*: sterile pyuria, with haematuria, pyrexia and malaise;
    (b) *meningitis*: slow, insidious onset, with high mortality;
    (c) *bone and joints*: rare, most commonly affects the lumbar spine;
    (d) *lymph glands*: the most common site of non-pulmonary tuberculosis, the cervical lymph nodes are most frequently involved and may be the only site of infection, particularly in children;
    (e) *abdominal tuberculosis*: rare, difficult to diagnose.

Laboratory diagnosis. Microscopy and culture of relevant specimens, depending on suspected site of infection, include sputum, bronchoscopy material, pleural fluid, urine, joint fluid, biopsy tissue and cerebrospinal fluid. Direct detection of *M. tuberculosis* DNA in specimen by the polymerase chain reaction (PCR) as a diagnostic method is under evaluation. Skin testing (Mantoux test) for diagnosis of active tuberculosis is only appropriate in countries where incidence of tuberculosis is low and BCG vaccination is not administered routinely (e.g. USA).

Treatment. This is often started on clinical suspicion; first-line agents include ethambutol, isoniazid, rifampicin and pyrazinamide. Combinations of up to four drugs are used to prevent emergence of resistance. Treatment regimens range from 6 months to 1 year. Resistance to first-line agents is increasing particularly in Asia and Africa, and, more recently, in areas of the USA in association with human immunodeficiency (HIV) infection. Second-line anti-mycobacterial agents include cycloserine and kanamycin.

Prevention and control. Strategies include improved living standards (housing, nutrition); early recognition of new cases followed by isolation of patients with open tuberculosis for at least 2 weeks from commencement of therapy; follow-up of contacts, with skin testing and radiography as appropriate; prophylaxis with rifampicin for close contacts, particularly non-immune children and the immunocompromised; immunization (BCG vaccine).

CHAPTER 34

Gastrointestinal Infections

Infective gastroenteritis. This is an inflammation of the gastrointestinal tract (GIT) caused by microorganisms. Spread is principally by the faecal–oral route, either directly or via vectors such as food or water. Symptoms include diarrhoea, vomiting, abdominal pain and fever.

Food poisoning. This is the illness resulting from consumption of food contaminated with pathogenic organisms and/or their toxins. It mainly affects GIT, but can affect other sites (e.g. *Listeria monocytogenes*, meningitis; *Clostridium botulinum*, paralysis).

This chapter outlines the important causes of infective gastroenteritis and food poisoning (Table 34.1).

FOOD POISONING AND GASTROENTERITIS

Organism	Incubation period (h)	Clinical features			
		Diarrhoea	Vomiting	Abdominal pain	Fever
Bacillus cereus	0.5–6.0	+M	+S	0	0
Campylobacter jejuni	48–120	+S	+M	+S	+
Clostridium perfringens	12–24	+M	0	+	0
Clostridium botulinum	12–36	0	±	0	0
Salmonellae	18–48	+M	±	+	0
Staphylococcus aureus	1–6	±	+S	0	0
Vibrio parahaemolyticus	6–36	+M	+M	+	0

0, absent; +, often present; ±, occasional; M, moderate; S, severe.

Table 34.1 Some common bacterial causes and clinical features of food poisoning and gastroenteritis.

BACTERIA

Shigella

The four species of shigella (*Shigella sonnei*, *S. flexneri*, *S. boydii* and *S. dysenteriae*) are important causes of bacillary dysentery, which is characterized by diarrhoea containing blood, mucus and pus.

Epidemiology. Spread is by faecal–oral route. A small infecting dose is required, therefore person-to-person spread is common. Food can act as a vector occasionally. It can also spread via fomites (inanimate objects that become contaminated by the organism), particularly in toilet areas. Humans are the only reservoir of infection. In the developing world, dysentery is normally caused by *S. dysenteriae* and *S. boydii*, and is common in both adults and children. In developed countries, *S. sonnei* and *S. flexneri* infections are more common and occur more in young children, particularly those associated with nurseries.

Pathogenesis. The organism invades the colonic mucosa, causing inflammation and ulcers, which result in mucosal bleeding. Blood stream invasion is rare. *S. dysenteriae* also produces an exotoxin which disrupts absorption, resulting in diarrhoea.

Clinical features. The incubation period of 1–4 days is followed by diarrhoea, typically with blood, mucus and pus, usually with abdominal pain and, infrequently, fever. Symptoms normally last 3–4 days. *S. dysenteriae* is associated with severe disease; infections with *S. sonnei* are often mild. Prolonged carriage for several months can occur, but long-term chronic carriage is not a feature of shigellosis.

Laboratory diagnosis. Stool culture on selective media.

Treatment and prevention. Antibiotic treatment (e.g. ciprofloxacin) is used for severe disease only as infections are usually self-limiting. Multi-resistant strains have appeared in some countries. Good personal hygiene and public health measures aid prevention.

Salmonella

ENTERITIS

Numerous salmonella species can cause bacterial food poisoning resulting in enteritis. In contrast, *Salmonella typhi* and *S. paratyphi* do not cause diarrhoeal disease, but invade the blood stream to cause a systemic infection (typhoid and paratyphoid fever).

Epidemiology. Faecal–oral spread is the main route of transmission, principally via food vectors and sometimes water; the infecting dose is high, therefore multiplication in food is necessary. Direct person-to-person spread may occur, mainly in hospitals. Salmonella are commonly carried by domestic and wild animals; (carriage rates in chickens can be >70%). Humans are the sole host for *S. typhi* and *S. paratyphi*. Poor hygiene practices in kitchens are important in the spread of salmonella, particularly where cross-contamination between raw meat and cooked food, and inadequate cooking or storage of food occurs. Following infection, a chronic carrier state may occasionally occur.

Pathogenesis. Salmonella multiply in the gut and cause direct mucosal damage; the overall increased fluid loss results in diarrhoea.

Clinical features. The incubation period is 18–48 h, occasionally longer. Diarrhoea (occasionally blood-stained), abdominal pain and fever are typical features. Vomiting may occur. Symptoms are self-limiting, lasting 2–3 days. A small number of cases become long-term carriers of salmonella. Septicaemia can occur in the elderly, neonates and the immunocompromised. Meningitis, osteomyelitis and septic arthritis are rare complications of invasive salmonellosis.

Laboratory diagnosis. By stool culture; blood culture in systemically ill patients.

Treatment and prevention. Salmonellosis is a self-limiting infection and antibiotic treatment is often unnecessary. Septicaemic patients, those at risk of invasive disease or patients with enteric fever should be treated with antibiotics (e.g. cotrimoxazole, ciprofloxacin). Prevention includes good kitchen practice (p. 317) and measures to reduce carriage of salmonella in livestock.

Campylobacter

Several species of campylobacter cause human infection; *Campylobacter jejuni* is the most frequently isolated species.

Epidemiology. First recognized in the early 1970s as a cause of human gastroenteritis, campylobacter is now the most common cause of bacterial gastroenteritis in the UK. There are a number of animal reservoirs, including cattle, sheep, poultry, pets and wild birds. Transmission is via the faecal–oral route, but person-to-person spread may occur. Ingestion of approximately 10^4 organisms is required for infection. Contaminated food and unpasteurized milk are important vectors. Long-term carriage is unusual.

Clinical features. The incubation period is normally 2–5 days, but may be prolonged. Diarrhoea occurs with frank blood, mucus and pus. Fever and pain are prominent features which may mimic other conditions, e.g. appendicitis. Symptoms last 5–7 days, but may be prolonged. Complications are rare, but septicaemia can occur, particularly in the immunocompromised.

Laboratory diagnosis. By isolation from stool specimens.

Treatment and prevention. Patients with protracted symptoms should be treated with erythromycin. Prevention is as for salmonella infection.

Vibrio cholera

Vibrio cholera is the cause of cholera, an intestinal infection characterized by profuse, watery diarrhoea.

Epidemiology. Cholera is endemic in many developing countries with local epidemics precipitated by natural disasters or wars. In developed countries cases may follow travel to endemic areas. Cholera is a disease of humans, with no animal reservoirs. Transmission is via water or food contaminated by human faeces; although direct person-to-person spread may occur. Asymptomatic excretion of the organism may occur.

Pathogenesis. The organism colonizes the mucosal surface of the intestine and produces an enterotoxin, resulting in loss of water and electrolytes.

Clinical features. Copious watery diarrhoea, up to 25 l/day (often described as 'rice water' stools); this may result in severe dehydration. Mortality is related to the dehydration and blood stream invasion is not a feature.

Laboratory diagnosis. By stool culture.

Treatment and prevention. Fluid and electrolyte replacement is the most important aspect of management; oral tetracycline reduces symptoms. Long-term prevention is by public health measures, in particular the separation of sewage from clean water supplies.

Escherichia coli

E. coli is part of the normal flora of the large intestine, but may cause enteritis by a number of distinct pathogenic mechanisms.

ENTEROPATHIC E. COLI (EPEC)

This was an important cause of outbreaks of diarrhoeal disease amongst infants but is now much less common. It is caused by a number of distinct *E. coli* 'O' serotypes (e.g. O111, O127) which cause direct damage to the intestinal villi. Diagnosis is by stool culture and *E. coli* serotyping.

ENTEROTOXOGENIC E. COLI (ETEC)

ETEC is an important cause of travellers' diarrhoea and diarrhoeal disease in children in developing countries; ETEC produce enterotoxins similar to cholera toxin, which result in profuse, watery diarrhoea. Laboratory diagnosis is by culture and demonstration of toxin production.

VEROTOXOGENIC E. COLI (VTEC)

VTEC cause two distinct conditions, *haemorrhagic colitis* (HC) and *haemolytic uraemic syndrome* (HUS). VTEC strains produce a toxin which may act locally on the gut mucosa, resulting in bloody diarrhoea (HC), or act systemically, resulting in haemolysis and renal failure (HUS). VTEC strains belong to a number of distinct 'O' serotypes, most commonly serotype O157.

VTEC are a cause of food poisoning and several large outbreaks of VTEC infections have been described associated with the consumption of meat products, particularly hamburgers. Laboratory diagnosis is by stool culture, serotyping of isolates and demonstration of toxin production; retrospective diagnosis can be made by serology.

ENTEROINVASIVE E. COLI (EIEC)

EIEC produce a shigella-like infection, with invasion of the intestinal mucosa, resulting in diarrhoea containing blood, pus and mucus.

Note: E. coli enteritis should only be treated with antibiotics (e.g. cefuroxime) if systemic infection is apparent.

Clostridium difficile

Epidemiology. C. difficile is part of the normal flora of the large intestine, particularly of children and the elderly. Person-to-person spread can occur, resulting in hospital outbreaks of *C. difficile* colitis.

Pathogenesis. Antibiotic treatment leads to disturbance of the normal gut flora and allows *C. difficile* to multiply. Some strains produce toxin A, which is responsible for the pathological change in the gut mucosa, and toxin B, a powerful cytotoxin which results in a cytopathic effect on tissue culture cells.

Clinical features. C. difficile causes a range of illnesses, from mild diarrhoea to pseudomembranous colitis characterized by severe bloody diarrhoea, systemic symptoms, and the formation of pseudomembranes in the colon. Colonic perforation may ensue.

Treatment and prevention. Oral vancomycin or metronidazole are the drugs of choice for treatment of *C. difficile* infection. Prevent by avoiding unnecessary use of broad-spectrum antibiotics and barrier nursing patients with the condition.

Clostridium perfringens

C. perfringens is part of the normal flora of the large intestine of animals and humans. It causes food poisoning and gas gangrene. The latter infections are often associated with peripheral ischaemia and can be caused by other clostridial species (e.g. *C. septicum, C. histolytica*); gas gangrene is covered in Chapter 29.

FOOD POISONING

Pathogenesis. Spores often contaminate meat and can survive cooking, particularly in large pieces of meat. If the food is then stored unchilled, the spores germinate and the organism multiplies. Following ingestion, *C. perfringens* produce an enterotoxin (α-toxin) resulting in diarrhoea.

Clinical features. The incubation period is 12–24 h, followed by abdominal pain and explosive diarrhoea; symptoms last only a short time (<6 h).

Laboratory diagnosis. By isolation of the organism from faecal and/or food samples.

Treatment and prevention. Supportive measures only are required for treatment. Good kitchen practices aid prevention (p. 317).

Staphylococcus aureus

S. aureus is an important cause of toxin-mediated food poisoning.

Pathogenesis. The organism contaminates food, typically via a food handler with a staphylococcal skin lesion. Contaminated food left at ambient temperature allows the organism to multiply and produce toxin, which can survive heating to 100°C for 30 min. Typical foods include cooked meat and cream cakes. Following ingestion, the toxin is absorbed rapidly and acts on the central nervous system, resulting in severe vomiting.

Clinical features. The incubation period is 1–6 h, followed by severe abdominal pain and vomiting; diarrhoea is not a common feature. The condition is self-limiting, lasting 12–24 h.

Laboratory diagnosis. By isolation of the organism from food and demonstration of its ability to produce enterotoxin; however, often no viable organisms are present in food. Direct detection of toxin in food can be made by immunoassays.

Treatment and prevention. Supportive therapy only; good kitchen practice for prevention (p. 317).

Yersinia enterocolitica

Y. enterocolitica is an uncommon cause of diarrhoeal disease, which may be food associated. The organism invades the terminal ileum and may result in a mesenteric adenitis, mimicking appendicitis.

Vibrio parahaemolyticus

A cause of food poisoning often associated with consumption of shellfish. Pain, fever and diarrhoea follow 12 h after ingestion.

Clostridium botulinum

This is a rare cause of food poisoning, more common in countries where there is small-scale canning of fruit and vegetables. *C. botulinum* spores contaminate the food and survive the heating process. The organism multiplies and toxin is produced. Pre-formed toxin is then ingested, absorbed from the gastrointestinal tract, and blocks neurotransmission of peripheral nerves, leading to a flaccid paralysis and eventually respiratory arrest. Symptoms usually appear within 12–36 h, occasionally several days, after eating contaminated food.

Laboratory diagnosis is dependent on detecting toxin, either in food or faeces by culture or serological tests. Treatment involves intensive care with respiratory support, and administration of anti-toxin. Preventive measures in the canning industry involve heating food to temperatures high enough to destroy spores.

VIRUSES

Viruses frequently result in gastroenteritis, particularly in children, and in the developing countries, they are a major cause of death. Up to 1 million infants worldwide die of viral gastroenteritis per year. Fluid replacement with oral electrolyte solutions is a critical part of management.

Rotavirus

These are morphologically characteristic viruses with a wheel-like appearance.

Epidemiology
- Many different serotypes exist. Different serotypes cause diarrhoeal disease in other mammals, including cats, dogs, cattle, sheep and pigs.
- The infecting dose is small and spread is generally person-to-person. Rotavirus infections are most common in the winter, with epidemics occurring occasionally in nurseries.

Pathogenesis. Damage to enterocyte transport mechanisms results in loss of water and electrolytes and diarrhoea.

Clinical features. The incubation period is 1–2 days. Vomiting and diarrhoea occur, often preceded by upper respiratory tract symptoms (cough, coryza). Infections are self-limiting and there is no specific treatment, apart from fluid replacement.

Laboratory diagnosis. By electronmicroscopy or the direct detection of virus particles by immunoassays (e.g. enzyme-linked immunosorbent assay (ELISA)).

Other viruses

Aetiology
- *Caliciviruses* (e.g. Norwalk agent) cause 'winter vomiting disease'.
- *Adenoviruses* (types 40 and 41) cause acute diarrhoea in young children.
- *Small round viruses* are associated with food poisoning (particularly seafood) and 'winter vomiting disease'.
- *Astroviruses* cause mild, self-limiting diarrhoea.

Laboratory diagnosis. By electronmicroscopy of faecal samples.

Treatment. There is no specific treatment, except fluid replacement. In small children in developing countries, fluid replacement with electrolyte solutions is an extremely important part of management.

PROTOZOA

Giardia lamblia

Epidemiology. G. lamblia has a worldwide distribution. It is spread by the faecal–oral route and is frequently associated with drinking water contaminated with cysts.

Pathogenesis. The life cycle of *Giardia* is described in Chapter 21. Trophozoites (Plate 34) attach to the mucosa of the small intestine: epithelial cells are damaged, affecting transport mechanisms, impairing absorption and resulting in diarrhoea.

Clinical features. The incubation period is 1–3 weeks. Symptoms include diarrhoea (loose, fatty, foul-smelling stools), mild abdominal pain and discomfort, and are self-limiting (approximately 7–10 days). Chronic infection may occur in compromised hosts.

Laboratory diagnosis. By microscopy of stool samples or duodenal aspirates.

Treatment and prevention. Treatment is with metronidazole. Public health measures are important in prevention to ensure clean drinking water.

Cryptosporidium parvum

Aetiology. Cryptosporidium has only recently been recognized as a cause of human infection. It was previously an established cause of diarrhoea in calves. It has a complex life cycle with asexual and sexual phases of development in the same host (see Chapter 21, p. 125).

Epidemiology. Transmission is by the ingestion of cysts acquired from either direct contact with farm animals or infected cases, or via contaminated water or milk. The cyst releases sporozoites in the small intestine, which invade epithelial cells where they form schizonts and merozoites. These reinvade other epithelial cells. A sexual phase occurs with oocysts forming.

Clinical features. Diarrhoea is normally mild and lasts several weeks. It can be severe in immunocompromised patients, particularly AIDS patients, and continue for months. An asymptomatic carrier state may occur.

Laboratory diagnosis. By microscopy of faecal samples stained by a modified acid-fast stain.

Treatment and prevention. There is no recognized effective treatment for cryptosporidial diarrhoea. Prevention includes care with personal hygiene when handling farm animals.

Entamoeba histolytica

Infections with *E. histolytica* are endemic in subtropical and tropical countries. The life cycle and associated infections are described in Chapter 21, p. 122.

CHAPTER 35

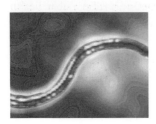

Liver and Biliary Tract Infections

The liver has an important role in removing microorganisms and their products from the circulation. It may be involved in infection as a primary site (e.g. viral hepatitis) or as part of a multisystem infection (e.g. miliary tuberculosis, metastatic abscesses). Liver damage (jaundice, deranged liver enzymes) may be a feature of severe sepsis. Patients with chronic liver disease also have lowered immune defences and are susceptible to infection.

The liver may become infected via the blood stream (the dual blood supply of the liver via the hepatic artery and the portal vein results in the liver being particularly vulnerable to blood-borne infection) or via the common bile duct, particularly when bile drainage is obstructed.

CHOLECYSTITIS AND CHOLANGITIS

Aetiology and pathogenesis. The incidence of cholecystitis and cholangitis increases with age and is more frequent in women. Obstruction of the bile duct results in stagnant bile which becomes infected with organisms from the intestinal tract. Infection may involve the bile duct (cholangitis) and/or the gall bladder (cholecystitis). Coliforms are the most common organisms isolated, particularly *Escherichia coli*, *Klebsiella* and *Proteus*; infections are frequently mixed and may include anaerobes.

Clinical features. There is right upper quadrant pain and tenderness, fever and malaise, and mild jaundice in some patients.

Complications. Gangrene and perforation of the gall bladder occurs in <10% of patients. Empyema of the gall bladder and pancreatitis are also rare complications.

Laboratory diagnosis. Imaging techniques and blood cultures are used for diagnosis.

Management. Intravenous antibiotics are given to cover aerobic Gram-negative bacilli and anaerobes. Surgery is required to relieve obstruction but is normally delayed until the acute infection is controlled.

LIVER ABSCESSES

Aetiology and pathogenesis. Liver abscesses may result from spread via the systemic circulation, the bile duct (cholangitis) or the portal vein (intra-abdominal sepsis). The most common organisms isolated are *Streptococcus milleri*, *E. coli*, often mixed with anaerobes (e.g. *Bacteroides fragilis*) and *Staphylococcus aureus*.

Clinical features and complications. Swinging fever is present; pain and tenderness are felt in the right upper quadrant.

Clinical and laboratory diagnosis. Ultrasonic and computerized tomographic (CT) scans are performed. Blood cultures (positive in about 30% of patients) and culture of aspirated pus are carried out.

Treatment. This includes surgical drainage and antibiotics (dependent upon culture results).

AMOEBIC ABSCESSES

Pathogenesis. These are a complication of intestinal amoebiasis caused by *Entamoeba histolytica* (Chapter 21, p. 122). Severe intestinal disease allows organisms to spread to the liver via the portal vein. The abscess increases in size and liver tissue is destroyed; amoebae are present at the margin of the abscess, the centre being filled with blood-stained pus. Abscesses are often single and in the right lobe of the liver.

Clinical features. Symptoms are often minimal; there may be a slight fever with right upper quadrant tenderness. Liver function tests may be slightly elevated.

Clinical and laboratory diagnosis. This is by examination of stool samples for amoebic trophozoites and cysts; microscopy of aspirated material from the abscess; serological tests (e.g. ELISA) for antibodies to *E. histolytica*; and CT scan or ultrasound of the liver.

Treatment. Metronidazole; surgical drainage is now rarely performed.

OTHER PARASITES

Hydatid cysts

The dog tapeworm, *Echinococcus granulosus*, may infect humans and result in hydatid cysts, particularly in the liver (p. 147).

Fasciola (liver fluke)

The adult worms of the liver fluke may result in biliary obstruction and liver cirrhosis (p. 148).

Schistosomes

Schistosomal eggs may lodge in the liver and result in liver fibrosis and subsequent portal hypertension (p. 151).

VIRAL HEPATITIS

Viruses that primarily affect the liver are described as hepatotropic viruses and currently five are recognized, hepatitis A, B, C, D and E.

Hepatitis A virus (HAV)

HAV is an RNA virus and is a member of the picornavirus family.

Epidemiology
- HAV is transmitted by the faecal–oral route. It is excreted in the faeces about 1 week before symptoms appear and up to 1 week after. Outbreaks have been associated with infected food handlers and ingestion of contaminated shellfish.
- Incidence is highest in areas of the world with poor sanitation, and is most common in children.

Clinical features
- An incubation period of 2–6 weeks is followed by malaise, anorexia, nausea and right upper quadrant pain.
- Jaundice appears during the second week and normally lasts several weeks, but may be prolonged. About one-quarter of patients are anicteric.
- Fulminant liver failure occurs in about 0.1% of cases. Chronic liver disease is not a feature of HAV infection.

Laboratory diagnosis
- Liver function tests show markedly elevated serum levels of liver enzymes (aminotransferases) and bilirubin.
- Diagnosis is confirmed by the detection of serum IgM antibodies to HAV. Measurement of anti-HAV IgG is used to confirm immunity to HAV (past infection or vaccination).

Treatment and prevention. There is no specific antiviral treatment; general measures include bed-rest; hospitalization is rarely necessary. Passive immunization with normal immunoglobulin provides protection for about 6 months for travellers to endemic areas; hepatitis A vaccine is now replacing the need for passive immunization.

Hepatitis B virus (HBV)

Hepatitis B is a DNA virus and is a member of the hepadnaviridae virus family.

Epidemiology
- Transmission of hepatitis B virus (HBV) is by percutaneous and permucosal routes. There are three important mechanisms of transmission:
    (a) *contact transmission* via bodily secretions (e.g. blood, semen and vaginal fluid);
    (b) *maternal–infant transmission* across the placenta or during delivery;
    (c) *percutaneous transmission*: at-risk groups include parenteral drug abusers and health care workers involved in needlestick injuries. Infection via transfusion of contaminated blood products is rare in most countries because of active screening of blood donors.
- Prevalence patterns of HBV infection vary. In developed countries prevalence is <1%, but in some endemic areas (e.g. North Africa, Asia) prevalence of the carrier state may be as high as 20%.
- In high-prevalence areas, maternal–infant transmission is important, whilst in areas of low prevalence, percutaneous and sexual routes are more important modes of transmission.

Clinical features and complications. An incubation period of 6 weeks–6 months is followed by malaise, anorexia and jaundice. Symptoms associated with immune complex disease occur, rarely, e.g. arthralgia, vasculitis, glomerulonephritis. Asymptomatic infections may occur. Complications include fulminant liver failure (1%) and chronic hepatitis (10%).

Laboratory diagnosis. Three HBV antigens and the corresponding antibodies are used in the diagnosis of acute and chronic HBV infection.
- Hepatitis B surface antigen (HBsAg) is the first serological marker of acute HBV infection, appearing several weeks before symptoms.
- Hepatitis Be antigen (HBeAg) appears soon after HBsAg and is the first antigen to disappear in patients who are recovering from hepatitis B; its presence is associated with increased infectivity.
- IgM antibodies to the hepatitis B core antigen are a useful marker of acute HBV infection in patients presenting after surface antigen is no longer detectable.
- Presence of antibody to HBsAg, the last serological marker to appear, indicates past infection with HBV or previous vaccination.

Treatment and prevention. There is no specific antiviral treatment for acute hepatitis B. Prevention is by: screening of blood donors and products; use of disposable needles and other instruments and efficient sterilization of re-usable medical instruments. A genetically engineered vaccine is now available and should be given to at-risk groups, particularly health care workers. Specific immunoglobulin (passive immunization) may be given to non-immunized persons exposed to HBV (e.g. in a needlestick injury) and to babies born to HBeAg positive carrier mothers.

CHRONIC HEPATITIS B

Approximately 10% of patients with acute hepatitis B will develop chronic disease. The incidence varies inversely with age; about 90% of neonates and <10% of adults will develop chronic hepatitis B.

Definition. The persistence of HBsAg >6 months after acute infection.

Clinical features and outcome. About 80% of cases have minimal liver damage, remain largely symptom-free and often undergo spontaneous remission (chronic persistent hepatitis). Approximately 20% of cases follow a more aggressive course (chronic active hepatitis); liver damage is progressive and may result in cirrhosis. There is an increased risk of hepatocellular carcinoma.

Laboratory diagnosis. Based on failure to clear HBsAg. HBeAg may also persist and indicates continued viral replication and increased infectivity.

Management. Patients should reduce alcohol intake, which may act as a co-factor in the development of cirrhosis; liver function tests should be monitored.

Hepatitis C virus (HCV)

Hepatitis C virus is an RNA virus related to the pestivirus genus of the flavivirus family.

Epidemiology
- HCV is the cause of about 90% of hepatitis cases previously known as 'non-A, non-B hepatitis' or 'non HBV-transfusion-related hepatitis'.
- Transmission occurs in a similar manner to HBV (infected blood products, intravenous drug abuse, sexual transmission).

Clinical features. The incubation period is 2–6 months. The hepatitis is often mild but frequently progresses to chronic active hepatitis and cirrhosis after many years (>25 years), with an increased risk of developing liver carcinoma.

Laboratory diagnosis. This is by serology, for the presence of antibody to HCV and polymerase chain reaction for the detection of HCV RNA.

Prevention. Methods are similar to HBV, including the screening of blood products. A vaccine has not been developed.

Hepatitis D virus (HDV)

The hepatitis D virus or delta agent is a defective RNA virus which can only replicate in HBV-infected cells. Transmission is via infected blood and sexual intercourse. HDV accentuates HBV infection, resulting in more severe liver disease. Laboratory diagnosis is by HDV antibody (or rarely antigen) detection. No vaccine is available at present.

Hepatitis E virus (HEV)

- This is a small RNA virus, spread by the faecal–oral route often via water; large waterborne epidemics have occurred in India. It is rare in the developed world.
- The incubation period is 6–8 weeks, followed by mild hepatitis; severe fulminant HEV hepatitis may occur in pregnant women.
- The virus is eliminated on recovery; chronic carriage of the HEV has not been described. Diagnosis is by serology.

CHAPTER 36

Urinary Tract Infections

URINARY TRACT INFECTIONS (UTI)

Laboratory definition. Significant bacteriuria is defined as the presence of $\geq 10^8$ bacteria per litre of a fresh midstream urine (MSU) sample; occasionally patients with UTI symptoms may have fewer organisms (10^6–10^7/litre).

Clinical definitions. There are three clinical syndromes associated with UTI.
1 *Frequency dysuria syndrome.* Dysuria and frequency due to:
   - bacterial cystitis with significant bacteriuria and often associated with pyuria and haematuria;
   - abacterial cystitis ('urethral syndrome'), where no microbial cause is identified. Some cases may be associated with organisms that are not cultured by routine laboratory methods.
2 *Acute bacterial pyelonephritis.* Infection of the kidney with symptoms of loin pain and tenderness, and pyrexia accompanied by bacteriuria and pyuria.
3 *Covert bacteriuria (asymptomatic bacteriuria).* This involves significant numbers of bacteria present in the urine of apparently healthy people, with no associated symptoms. It is important in childhood and pregnancy.

Aetiology. The bacteria that commonly cause UTI are commensals of the perineum or lower intestine (Table 36.1). *Escherichia coli* is the most common cause. *Proteus mirabilis* is associated with urinary stones. Other Enterobacteriaceae (e.g. *Klebsiella*, *Enterobacter* and *Serratia*) are commonly associated with hospital-acquired UTI. *Staphylococcus saprophyticus* is associated with UTI in sexually active young women. Non-bacterial causes of UTI are shown in Table 36.2.

Immune defence of the urinary tract
- Hydrodynamic forces: the flow of urine removes organisms from the bladder and urethra.
- Phagocytosis by polymorphs on the bladder surface.

BACTERIAL CAUSES OF UTI

Organism	Community-acquired (%)	Hospital-acquired (%)
Escherichia coli	75	40
Staphylococcus saprophyticus	5	1
Staphylococcus epidermidis	2	3
Proteus mirabilis	3	10
Enterococcus faecalis	5	8
Other coliforms	5	25
Pseudomonas aeruginosa	1	5
Candida albicans	4	8

Table 36.1 Common bacterial causes of UTI (approximate %).

NON-BACTERIAL CAUSES OF UTI

Group	Organism	Infection
Viruses	Adenoviruses	Haemorrhagic cystitis
	Human polyoma virus	Infections in kidney and ureter
Parasites	Trichomonas vaginalis	Urethritis
	Schistoma haematobium	Bladder inflammation
Fungi	Candida albicans	UTI in immunocompromised patients

Table 36.2 Non-bacterial causes of UTI.

- Presence of IgA antibody on the bladder wall.
- Mucin layer on the bladder wall preventing bacterial adherence.
- Urinary pH.

Bacterial virulence factors include:
- Fimbriae: certain serotypes of E. coli have specific fimbriae (pili) which facilitate colonization and adherence to the periurethral areas, the urethra and bladder wall.
- Capsules: some strains of E. coli produce a polysaccharide capsule which inhibits phagocytosis and is associated with the development of pyelonephritis.

Patient factors include:
- Short urethra in females: sexual intercourse facilitates the passage of organisms up the urethra (honeymoon cystitis).

- Increasing age.
- Structural abnormalities causing outflow obstruction (e.g. prostatic enlargement; pregnancy; tumour); neurogenic bladder (e.g. in paraplegia); these result in residual urine in the bladder, which can act as a nidus of infection. Reflux of urine from the bladder up the ureter (vesico-ureteral reflux) into the kidney can result from anatomical abnormalities and cause pyelonephritis. Renal stones can be associated with pyelonephritis.
- Diabetes mellitus; immunosuppression (e.g. steroids, cytotoxic drugs).
- Instrumentation (e.g. surgery or the use of urinary catheters); bacteria may be introduced into the bladder on catheter insertion, or may grow up the catheter into the bladder at a later stage.

Clinical features
- *Lower UTI (cystitis, urethritis)*: Frequency of micturition, dysuria (pain on passing urine), and urgency (urgent need to micturate); fever. In the elderly and infants, UTI may be asymptomatic or have non-specific presentations, e.g. confusion or unsteadiness (elderly), poor weight gain, irritability (infants).
- *Upper UTI (pyelonephritis)*: Loin pain or tenderness and high fever, associated with symptoms of lower UTI.

Complications
Septicaemia, particularly in the elderly. Persistent or recurrent UTI in the young can result in renal scarring and is an important cause of chronic renal failure. Asymptomatic bacteriuria in pregnancy is associated with pre-term delivery and low birth weight.

Laboratory diagnosis
- *Specimen collection*: Avoid contamination from perineal bacteria by collecting clean-catch midstream samples; in infants, adhesive bags may be used, but supra-pubic aspiration may be necessary. In catheterized patients, fresh urine should be collected directly from the catheter via a syringe and needle and not from the drainage bag. Investigation for renal tuberculosis requires the collection of complete urines on three consecutive mornings.

Urine specimens should be transported to the laboratory within 2 h to avoid bacterial multiplication and a false diagnosis of UTI; if delayed, samples should be refrigerated at 4°C or the dipslide method used (an agar-coated slide is dipped in the patient's urine and then incubated).
- *Laboratory analysis*: Microscopy: urine is examined for the presence of leucocytes ($>10/mm^3$ indicates significant pyuria in patients without a

catheter). Culture: various techniques are used to quantify bacterial numbers in urine samples.

Interpretation of laboratory results
- $\geq 10^8$ bacteria/l; significant bacteriuria.
- 10^7–10^8 bacteria/l; may indicate infection; correlate with clinical factors such as the presence of symptoms or the use of antibiotics which might suppress bacterial growth.
- $<10^7$ bacteria/l, patient not receiving antibiotics; infection unlikely; if patient is symptomatic, suggests abacterial cystitis.
- Contamination of the specimen is characterized by growth of two or more bacterial species; a count $<10^7$ bacteria/l; and the presence of squamous epithelial cells from the periurethra.
- Catheter specimens: the presence of pyuria and bacteriuria is common and does not always indicate infection.
- Sterile pyuria: consider renal tuberculosis; other uncommon bacterial causes (e.g. mycoplasma); concurrent antibiotic treatment; and non-infective causes (e.g. foreign bodies, tumours).

Management
- *General measures*: encourage intake of increased volumes of fluid to flush out infected urine; double micturition to ensure bladder is empty and micturate last thing at night.
- *Antimicrobial agents*: first-line agents for use in uncomplicated community-acquired UTI include ampicillin, trimethoprim, nalidixic acid or nitrofurantoin. Complicated UTI with pyelonephritis often requires systemic antimicrobials.

Hospital-acquired UTI may be associated with resistant organisms and the choice of therapy is dependent on antibiotic sensitivity tests. Pregnant women with asymptomatic bacteriuria require antibiotic treatment.

Patients (particularly children) with recurrent UTI may require investigations for anatomical defects of the urinary tract or other underlying diseases (e.g. diabetes). Low-dose prophylactic antibiotics may be necessary.

Duration of therapy depends on the nature and severity of infection; for uncomplicated UTI, short doses of antibiotics (2–5 days) can be given and there is evidence that a single dose is effective. If symptoms persist, urine samples should be repeated 3 days after stopping the antibiotics. Longer courses of antibiotics are required for complicated UTI.

CHAPTER 37

Genital Infections (Including Sexually Transmitted Diseases)

Most infections of the genital tract are spread by sexual contact; causative organisms are generally exclusive human pathogens and survive poorly outside the host. Sexually transmitted diseases (STD; Table 37.1) are an important cause of morbidity and mortality, and the incidence worldwide is increasing; in some developed countries the incidence of certain STD (e.g. syphilis), has decreased because of active intervention programmes.

GONORRHOEA

Aetiology and epidemiology
- This disease is caused by *Neisseria gonorrhoea*, a Gram-negative diplococcus and an obligate pathogen of humans. Asymptomatic females act as a reservoir.
- The organism rapidly dies on drying, and direct contact is required for transmission; spread is via sexual contact or vertically from mother to neonate at birth. Gonorrhoea is an important STD worldwide, although the incidence has decreased in many developed countries.

Clinical features. The incubation period is approximately 2 days.
- Clinical features in the male are:

SEXUALLY TRANSMITTED DISEASES

Organism	Associated diseases
Neisseria gonorrhoea	Gonorrhoea
Human immunodeficiency virus (HIV)	AIDS
Chlamydia trachomatis	Non-specific urethritis
Candida albicans	Vaginal thrush, balanitis
Herpes simplex virus	Genital herpes
Treponema pallidum	Syphilis
Trichomonas vaginalis	Urethritis, vaginitis
Papillomaviruses	Genital warts

Table 37.1 Common sexually transmitted diseases.

(a) urethritis: purulent urethral discharge (Plate 59) and dysuria;
(b) rectal: associated with homosexuals, most cases are asymptomatic, but may cause proctitis (discharge and tenesmus);
(c) pharyngitis: associated with oral–genital sex;
(d) complications: prostatitis; epididymo-orchitis; bacteraemia with arthritis and skin lesions (e.g. pustules, Plate 60).
- Clinical features in the female are:
(a) infection of the endocervix, urethra and rectum, but symptoms (dysuria, vaginal discharge) may be mild or absent;
(b) complications: Bartholinitis (infection of Bartholin's glands); pelvic inflammatory disease (infection of salpinges with acute abdominal pain and pyrexia); bacteraemia with arthritis and skin lesions.
- Neonatal ophthalmia (see p. 287).

Laboratory diagnosis. This is by Gram stain of smears for Gram-negative diplococci, often within pus cells (intracellular), and culture.

Management
- High-dose intramuscular procaine penicillin; β-lactamase-producing strains can be treated with spectinomycin, cephalosporins (e.g. cefotaxime) or a quinolone antibiotic.
- Contact tracing and investigation of sexual partners should be carried out.

NON-SPECIFIC URETHRITIS (NSU)

Definition. Urethritis in the absence of gonococcal infection.

Aetiology and epidemiology. This is a common STD mainly in males, with more than double the incidence of gonorrhoea. Causative organisms include *Chlamydia trachomatis* (>50% cases), *Ureaplasma urealyticum* and *Trichomonas vaginalis*; other organisms (e.g. group B streptococci, *Gardnerella vaginalis* (see Plate 21), yeasts), may be associated with NSU occasionally. Mixed infections occur.

Clinical features. These are: in male patients, mucopurulent urethral discharge, often difficult to distinguish clinically from gonorrhoea; in female patients, usually asymptomatic, but may present with urethritis or cervicitis.

Laboratory diagnosis and treatment. See individual infections below.

Chlamydial infections

Aetiology and epidemiology. C. trachomatis; serotypes D–K; transmitted during sexual intercourse. Vertical transmission to neonate at delivery.

Clinical features. These are: in males, NSU; in females, cervicitis, NSU; may be asymptomatic in both males and females.

Complications
- Males: prostatitis, epididymitis, Reiter's syndrome (urethritis, arthritis, conjunctivitis).
- Females: Bartholinitis, pelvic inflammatory disease.
- Neonates: conjunctivitis (see p. 287).

Laboratory diagnosis. This is by: direct detection in urethral, cervical, or urine samples by staining smears with labelled (fluorescent; enzyme-linked (ELISA)) monoclonal antibodies; growth in tissue culture cell lines; and polymerase chain reaction assays.

Management. Tetracycline or erythromycin are given. Contact tracing and investigation of sexual partners should be carried out.

Trichomonas infection

Aetiology and epidemiology. T. vaginalis, a protozoan parasite, transmitted during sexual intercourse.

Clinical features
- Female patients: vaginitis with copious, foul-smelling discharge.
- Male patients: normally asymptomatic, they act as a reservoir of infection; may experience urethritis.

Laboratory diagnosis. Microscopy of a wet film of vaginal discharge shows motile trichomonads. Culture in special media can also be carried out.

Management. Treatment is with metronidazole; the sexual partner should also be treated.

SYPHILIS

Aetiology and epidemiology. This disease is caused by *Treponema pallidum*, a spirochaete and an obligate human pathogen. Syphilis has a worldwide distribution but is now uncommon in most developed countries (<1000 cases/year in UK). Transmission is by sexual contact and vertical spread via placenta to fetus.

Clinical features. Incubation is usually 14–21 days (10–90 days range). Several clinical stages of untreated syphilis are recognized.
- *Primary syphilis*: a painless ulcer (chancre), develops on the genitalia, perianal area, or, occasionally, at other sites. Treponemes multiply at the chancre and in the regional lymph nodes which become enlarged. Lesions heal over several months.
- *Secondary syphilis*: this occurs up to 2 months following the first stage. It is associated with: pyrexia, malaise, sore throat; a widespread macular-papular rash and lymphadenopathy; wart-like lesions (condylomata lata) may occur in perianal, vulva or scrotal areas; and oral and pharyngeal ulcers.
- Lesions heal and a dormant phase follows, *latent syphilis*, which may last up to 30 years before tertiary syphilis develops. Dormant treponemes may be present in the liver and spleen.
- *Tertiary syphilis*: treponemes start to multiply, resulting in granulomatous lesions (gummas) in various sites, including skin, subcutaneous tissues, mucous membranes, bones and joints. Cardiovascular complications include aortic valve incompetence, aortic aneurysm; central nervous system complications (neurosyphilis) include asymptomatic meningitis, tabes dorsalis (spinal cord damage), and general paralysis of the insane (cerebral damage).
- *Congenital syphilis*: spread via transplacental route; may result in intrauterine death or disease, presenting with congenital malformations at birth or several years later. Features include skin lesions, lymphadenopathy, failure to thrive, mental deficiency, peg-shaped teeth (Hutchinson's incisors), and bone and cartilage destruction (e.g. saddle-shaped nose).

Laboratory diagnosis. Dark-ground microscopy of exudates collected from lesions; serology.

Treatment and prevention. Treatment is with penicillin (usually a single injection of long-acting formulation) or tetracycline in penicillin-allergic patients. Prevention depends on adequate treatment, safe sex, contact tracing, screening and treating sexual partners. Congenital syphilis is prevented by serological screening in early pregnancy.

NON-SPECIFIC VAGINOSIS (NSV)

NSV is a vaginal infection with no specific aetiological agent. Characteristic features of NSV include: watery, foul-smelling ('fishy') discharge (no pus cells); and absence of pruritus and pain. Microscopy of discharge demonstrates the presence of large numbers of Gram-negative rods, often adherent to vaginal epithelial cells ('clue cells') (see Plate 21).

A number of bacteria are associated with NSV, including *Gardnerella vaginalis* and *Prevotella* (see Plate 22). Treatment is with metronidazole.

CANDIDIASIS

Candida albicans causes vulvovaginitis (vaginal thrush) in female patients. Low numbers of the organism are present as part of the vaginal commensal flora in some women; factors such as antibiotic therapy, steroid therapy (including the contraceptive pill), pregnancy and immunosuppression may disturb the normal flora and result in overgrowth of *C. albicans*. Clinical features include pruritus and a creamy vaginal discharge. Male patients become infected via sexual contact with infected females; the resulting infection involves the glans (balanitis), with inflammation and white plaques.

Laboratory diagnosis is by microscopy and culture of appropriate specimens. Treatment includes topical antifungals (e.g. nystatin) or oral fluconazole.

GRANULOMA INGUINALE

This genital infection is caused by *Calymmatobacterium granulomatis* and is usually found in the tropics. Sexual transmission may play a role in spread. Clinical features include papules and ulcers on the genitalia. Diagnosis is by microscopy of stained smears or biopsies of lesions. Treatment is with tetracycline.

CHANCROID

This is a genital infection characterized by painful ulcers with lymphadenopathy and caused by *Haemophilus ducreyi*. It has a worldwide distribution but occurs principally in tropical countries.

Diagnosis is by microscopy of Gram-stained smears of aspirates of ulcer or lymph nodes. Culture requires enriched media and prolonged incubation. Treatment is with erythromycin.

LYMPHOGRANULOMA VENEREUM

This infection is characterized by a genital ulcer with inguinal lymph node enlargement and suppuration. It is caused by *Chlamydia trachomatis* (serotypes L1, L2 and L3) and has a worldwide distribution, but is more common in tropical countries.

Dissemination of the organism may result in complications, e.g. proctitis, hepatitis, arthritis and meningoencephalitis. Chronic infection may result in genital strictures and fibrosis of lymphatics, with subsequent blockage and genital elephantiasis.

Diagnosis is by serology and treatment is with tetracycline.

GENITAL HERPES

Aetiology and epidemiology. STD caused by herpes simplex virus (HSV) type 2; type 1 strains are sometimes involved. It has worldwide distribution.

Clinical features
- Primarily, infection has an incubation period of 3–7 days. Clusters of vesicles appear on genitalia, progressing to painful, shallow ulcers. There is local lymphadenopathy.
- Resolution occurs in around 10 days, but HSV becomes latent in dorsal root ganglia; various stimuli (e.g. hormonal changes) may result in virus reactivation and recurrent lesions (genital cold sores).
- In pregnancy, active HSV infection at term may result in transmission to the neonate, with invasive HSV infection (e.g. encephalitis; see pp. 266, 288).

Laboratory diagnosis. Vesicle fluid or ulcer swabs are taken for virus isolation or direct detection of virus by immunofluorescence techniques or ELISA.

Treatment. Acyclovir (topical or systemic) reduces symptoms and infectivity.

GENITAL WARTS

This is a sexually transmitted infection (caused by human papillomaviruses). Following an incubation period of several months, warts appear on the genitalia, particularly at mucocutaneous junctions. It is associated with cervical carcinoma. Treatment is with podophyllin or by laser treatment.

CHAPTER 38

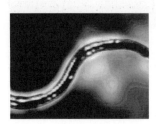

Infections of the Central Nervous System

Meningitis is defined as infection of the meninges, the membranous covering of the brain and spinal cord. *Encephalitis* involves direct invasion of the brain tissue, but is frequently accompanied by inflammation of the meninges (meningoencephalitis).

Infection may result from haematogenous spread (e.g. *Neisseria meningitidis*), invasion via nerves (e.g. Herpes simplex virus) or direct spread from a local focus of infection (e.g. mastoiditis).

Infections of the central nervous system may result in long-term sequelae (e.g. cranial nerve damage, epilepsy, deafness) as inflammatory changes and/or raised intracranial pressure can cause irreversible damage to neuronal tissue.

BACTERIAL MENINGITIS

The principal causes of bacterial meningitis are shown in Table 38.1.

Major causes of bacterial meningitis

NEISSERIA MENINGITIDIS

- Upper respiratory tract colonization occurs in up to 10% of healthy individuals; this may increase during epidemics. Spread is by the respiratory route (droplets). Invasion of the meninges follows bacteraemia/septicaemia.
- There is a small increase in risk of infection amongst close contacts of patients with meningococcal meningitis.
- Infections occur principally in children and young adults. In the USA and Europe, cases of meningococcal meningitis are sporadic and caused mainly by serotype B; epidemics of *N. meningitidis* serotypes A and C occur in Africa and South America.

Clinical features
- Normally of rapid onset, the principal symptoms are headache, fever and meningism (neck stiffness, irritability, photophobia). Concurrent sep-

BACTERIAL MENINGITIS	
Organism	Age group
Neisseria meningitidis	Children/young adults
Haemophilus influenzae (type b)	Children <5 years
Streptococcus pneumoniae	Elderly
	Children <2 years
β-haemolytic streptococcus group B	Neonates
Escherichia coli	Neonates
Listeria monocytogenes	Neonates/immunocompromised

Table 38.1 Causative organisms of bacterial meningitis according to age group.

ticaemia frequently leads to a typical petechial rash, which may become haemorrhagic.
- In fulminant cases, endotoxin release results in shock, with disseminated intravascular coagulopathy and multi-organ failure. In a few cases, haemorrhagic necrosis of the adrenal glands and intracranial bleeding occur, which is invariably fatal (Waterhouse–Friderichsen syndrome).

Prophylaxis and prevention
- In the developed world epidemics are rare; however, related cases in families or institutions can occur and chemoprophylaxis with rifampicin is recommended for close contacts.
- Health care workers in contact with cases are not at increased risk, but chemoprophylaxis is recommended for those involved in mouth-to-mouth resuscitation.
- A vaccine is available for protection against serotypes A and C, and is recommended for prophylaxis of close contacts of meningococcal cases caused by these serotypes.

Treatment. This is with intravenous penicillin; cefotaxime or chloramphenicol may be used in penicillin-allergic patients. Rifampicin is used to clear nasopharyngeal carriage.

HAEMOPHILUS INFLUENZAE TYPE B

- This occurs principally in children <5 years. Spread is by the respiratory route and symptomless upper respiratory colonization is common.
- Cases are often sporadic, but small outbreaks, particularly in nurseries, may occur.

Clinical features. It is similar to meningococcal meningitis except that onset is often less acute (1–2 days) and, although a petechial rash can occur, rarely, it is much less common.

Treatment and prophylaxis. Antibiotic therapy with cefotaxime and chloramphenicol. Rifampicin is used for chemoprophylaxis of close contacts. A vaccine is available and is part of immunization schedules in some countries.

STREPTOCOCCUS PNEUMONIAE

- Infection is most common in the young (<2 years of age) and the elderly; asplenic patients (e.g. sickle-cell disease) are at particular risk.
- Meningitis may be secondary to pneumococcal septicaemia or skull fracture (direct spread).
- Pneumococci are common commensals of the upper respiratory tract. Person-to-person spread is not a feature of pneumococcal meningitis.

Clinical features. Similar to other causes, but *S. pneumoniae* infection may be acute or insidious in presentation. Complications (e.g. deafness) are more common than with other bacterial causes.

Treatment. Until recently, pneumococci have remained sensitive to penicillin and erythromycin. However, penicillin-resistant strains have now emerged and are a particular problem in South Africa and parts of Europe. In the UK, up to 3% are resistant to penicillin. In these cases, cefotaxime is often used.

Prophylaxis and prevention. Contacts of cases of pneumococcal meningitis are not at an increased risk and chemoprophylaxis is inappropriate. A vaccine against many of the pneumococcal serotypes is available and is recommended for splenectomized and other immunocompromised patients.

Laboratory diagnosis of bacterial meningitis

Meningitis is a medical emergency and laboratory investigations are urgent.

- A lumbar puncture (LP) for cerebrospinal fluid (CSF) is carried out after raised intracranial pressure has been excluded. The typical features of CSF microscopy in bacterial meningitis are a raised polymorph count, a low glucose and an increased protein content (Table 38.2). Gram stain result may indicate the causative organism; culture is important for confirmation.

CEREBROSPINAL FLUID CHANGES

	Normal	Meningitis		
		Bacterial	Tuberculous	Viral
Leucocytes (cells/µl)	<5	500–20 000	200–1000	50–500
Polymorphs (%)		90	20*	5*
Lymphocytes (%)		10	80	95
Protein (g/l)	0.2–0.45	↑↑	↑↑↑	↑
Glucose (mmol/l)	4.5–8.5†	↓↓	↓↓	Normal

*Polymorphs may predominate in the early stages.
†About 80% of blood glucose level.

Table 38.2 Typical cerebrospinal fluid changes in meningitis.

- Assays for *N. meningitidis*, *H. influenzae* type b and *S. pneumoniae* antigens in CSF and urine can provide an early diagnosis, and may be useful in patients who have received antibiotics prior to LP.
- Blood cultures are positive in about 40% of cases of bacterial meningitis.
- Serology may give a retrospective diagnosis of meningococcal meningitis.

Treatment of bacterial meningitis

The fundamental principle in the treatment of bacterial meningitis is early, parenteral antibiotics in high doses. Intramuscular penicillin administered early by primary care physicians has been shown to reduce morbidity and mortality from meningococcal meningitis. Antibiotic therapy for common causes of bacterial meningitis in children and adults is shown in Table 38.3. There is evidence that concomitant steroid therapy reduces morbidity in children with meningitis.

NEONATAL MENINGITIS

Aetiology. *E. coli*, Group B β-haemolytic streptococci and *L. monocytogenes* are important causes of meningitis in the neonatal period; the three main causes of meningitis in children and adults are uncommon during the neonatal period.

Epidemiology. Acquisition is normally from the maternal genital or alimentary tract at or around the time of delivery.

ANTIMICROBIAL THERAPY OF ACUTE BACTERIAL MENINGITIS	
Organism	Treatment
*Aetiology unknown	Benzylpenicillin; or cefotaxime or chloramphenicol
N. meningitidis	Benzylpenicillin (chloramphenicol or cefotaxime if penicillin-allergic)
H. influenzae type b	Cefotaxime or chloramphenicol
S. pneumoniae	Benzylpenicillin (cefotaxime or chloramphenicol if penicillin-allergic or penicillin-resistant)
Note: Rifampicin is given for 48 h to patients with meningitis caused by N. meningitidis and H. influenzae to eliminate nasopharyngeal carriage. * Empirical treatment will be directed by the most likely causative organism and local antimicrobial sensitivity pattern.	

Table 38.3 Antimicrobial therapy for common causes of bacterial meningitis.

- E. coli is a common gut commensal and frequently contaminates the perineal area.
- Group B streptococci are part of the vaginal and perineal flora of about 30% of mothers.
- L. monocytogenes is a gut commensal in a small proportion of healthy mothers.

Neonates with prolonged hospital stay may acquire these organisms from nosocomial sources. Prematurity, low birth weight and prolonged ruptured membranes are important risk factors for neonatal meningitis.

Clinical features. Classical features (fever, meningism) are often absent. Presenting signs are subtle and include temperature instability, lethargy, decreased feeding and irritability.

Treatment. Early empirical therapy is imperative. Penicillin plus an aminoglycoside (e.g. gentamicin) or third generation cephalosporins are commonly used.

POST-OPERATIVE MENINGITIS

Meningitis is a serious complication of neurosurgical procedures. Gram-negative bacilli (*E. coli*, *Pseudomonas*) are important pathogens. Staphylococcus epidermidis is a frequent cause of cerebrospinal fluid (CSF) shunt infections. The wide range of possible pathogens requires broad-spectrum empirical therapy (e.g. ceftazidime). Vancomycin is often required for the treatment of S. epidermidis shunt infections.

TUBERCULOUS MENINGITIS

- Mycobacterium tuberculosis is a rare cause of bacterial meningitis in developed countries, but remains an important complication of tuberculosis in many areas of the world. It is a complication of either primary tuberculosis (normally pulmonary) or post-primary reactivation. Presentation is often indolent, with headache, malaise, anorexia, photophobia, neck stiffness and decreasing level of consciousness.
- Microscopy of CSF shows a lymphocytosis, high protein count and low CSF glucose (Table 38.2). A Ziehl–Neelsen stain of the CSF may show acid-fast bacilli, but definitive diagnosis is often reliant on CSF culture.
- Treatment follows the principles of anti-tuberculous triple therapy (rifampicin, isoniazid and pyrazinamide).
- Long-term complications are frequent because of the dense fibrous exudate; steroids appear to reduce complications.

VIRAL MENINGITIS

Viral meningitis, the most common form of meningitis, is often mild and full recovery is usual.

Aetiology. Important causes are enteroviruses (echoviruses, coxsackieviruses, polioviruses) mumps, and herpes simplex.

Clinical features. Influenza-like illness followed by meningism (neck stiffness, headache and photophobia).

Laboratory diagnosis. This is by: CSF microscopy which shows a lymphocytosis; Table 38.2; viral culture of CSF, stool and throat washings.

Treatment. There is no specific antiviral therapy, except acyclovir for herpes simplex.

FUNGAL MENINGITIS

Fungal meningitis occurs principally in immunocompromised patients, but can occur rarely in immunocompetent patients. Pathogens include *Cryptococcus neoformans*, *Aspergillus* species and *Candida* species.

PROTOZOAN MENINGITIS

Very rarely, the free-living amoeba, *Naegleria fowleri*, can cause meningitis. Patients normally have a history of swimming in warm, brackish water 1–2 weeks before presentation. Direct invasion of the meninges via the cribriform plate results in severe meningitis with a high mortality. Amoebae may be seen in unstained CSF preparations. Treatment is with amphotericin.

ENCEPHALITIS

Encephalitis is mainly caused by viruses, but other organisms are occasionally responsible (Table 38.4). Clinical signs include fever and vomiting, followed by decreased level of consciousness, focal neurological signs and eventually coma. Some important causes are outlined below; other causes are described in chapters dealing with individual pathogens.

Herpes simplex virus (HSV)

- Encephalitis is a rare complication of HSV infection. The incidence is increased in immunocompromised patients, neonates and the elderly.
- HSV may complicate primary (infancy) infection or follow viral reactivation (adults). Invasion of the temporal lobe is a prominent feature. Clinical diagnosis is often difficult; electro-encephalographic (EEG) studies and computerized tomography may be helpful.

Laboratory diagnosis. CSF may contain a few lymphocytes, but culture for HSV is often negative. The determination of serum antibodies is often unhelpful but detection of CSF antibodies is probably the most useful diagnostic test. The polymerase chain reaction has been introduced recently as a diagnostic test for HSV encephalitis. Brain biopsy is the definitive diagnostic test, but is difficult to justify.

Treatment and prevention. Treatment is with acyclovir. Neonatal HSV encephalitis is normally acquired from the maternal genital tract and pregnant women with active cervical herpes at term should undergo Caesarean section. Immunocompromised patients, particularly transplant patients, often receive acyclovir prophylaxis to prevent invasive HSV disease.

ENCEPHALITIS AND MENINGOENCEPHALITIS

Cause	Comments
Viruses:	
Herpes simplex virus	Most common cause of viral encephalitis
Mumps virus	Rare complication of mumps parotitis
Varicella-zoster virus	Complication of ophthalmic zoster
Cytomegalovirus	Immunosuppressed patients (e.g. AIDS, transplant patients)
Polio- and other enteroviruses	Important cause in developing world
Measles virus	Rarely in acute measles; reactivation of virus can occur years later, causing sub-acute sclerosing panencephalitis (SSPE)
Rabies virus	Zoonosis
Togaviruses	Arthropod-borne from various reservoirs
Retroviruses HTLVI / HIV	Endemic in some areas of the world; must be distinguished from other causes in AIDS patients
Bacteria:	
Treponema pallidum	Tertiary syphilis
Mycoplasma pneumoniae	Rare complication of mycoplasma pneumonia
Borrelia burgdorferi	Part of multisystem infection (Lyme disease)
Protozoa and fungi:	
Toxoplasma gondii	Immunocompromised
Cryptococcus neoformans	Meningoencephalitis, mainly immunocompromised
Plasmodium falciparum	Cerebral malaria
Trypanosomes	Larvae in brain; granulomas form
Toxocara	

Table 38.4 Principal causes of encephalitis and meningoencephalitis.

Poliovirus

- Transmission of the virus (an enterovirus) is via the faecal–oral route. The initial infection is often subclinical and central nervous system disease occurs in <1%. Fever and upper respiratory tract symptoms are followed by meningism. The virus invades the anterior horn cells, causing flaccid paralysis.
- Laboratory diagnosis is by virus isolation from faeces, throat washings or CSF.
- Prevention: both a killed and a live attenuated vaccine are available, and

vaccination programmes have led to a dramatic fall in cases in developed countries.

BRAIN ABSCESSES

Aetiology. Causative organisms related to pathogenesis; important causes include S. *aureus*, *Streptococcus milleri* and mixed aerobic/anaerobic infections (e.g. *E. coli* plus *Bacteroides*).

Pathogenesis. May follow trauma, surgery, septic emboli or direct invasion from adjacent infection (e.g. mastoiditis).

Diagnosis. Clinical and radiological; blood cultures or culture of aspirate may define the causative organism(s).

Treatment. Drainage and antibiotics as directed by culture.

ENCEPHALOPATHY (ASSOCIATED WITH SCRAPIE-LIKE AGENTS)

- The nature of the infectious agent is unclear and method of replication uncertain. Infectivity resists heat and chemicals, including formaldehyde.
- The infectious agent infects a range of mammals, including humans and replicate slowly with a long incubation period.
- Diagnosis is from histological changes in the brain. No treatment or vaccine is available.

Creutzfeld–Jakob disease (CJD)

This is a rare encephalopathy of humans with an uncertain mode of transmission. Transmission has occurred with corneal grafts, neurosurgery and growth hormone preparations. In some cases there may be a possible link to 'mad cow disease' (bovine spongiform encephalopathy (BSE)). Not destroyed by normal sterilization cycles in autoclaves. Requires 134°C for 18 min or exposure to sodium hydroxide.

Kuru

Transmission is from human to human by cannibalism and recorded mainly in Papua New Guinea.

TETANUS

- Tetanus toxin is produced by *Clostridium tetani* in infected wounds. The toxin is carried to the peripheral nerve axons and CNS, where it blocks inhibition of spinal synapses, resulting in overactivity of motor neurones with exaggerated reflexes, muscle spasms, lock-jaw (trismus), neck stiffness and opisthotonos (spasm of muscle of back, causing arching of trunk).
- Treatment is with anti-tetanus immunoglobulin, penicillin and by excision of the wound. Respiratory support is needed and mortality is high.
- Prevention is by immunization with toxoid vaccine.

CHAPTER 39

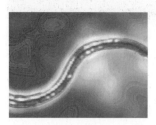

Septicaemia and Bacteraemia

DEFINITIONS

Bacteraemia. The transient presence of bacteria in the blood stream; clinical signs of septicaemia are absent, although there may be symptoms and signs of local infection which represent the source of most bacteraemias.

Septicaemia: The presence of bacteria in the blood stream associated with signs of shock, including tachycardia and reduced blood pressure. Symptoms and signs of localized infection, e.g. pneumonia, osteomyelitis, may also be present.

Endotoxic shock: Lipopolysaccharide (LPS ≡ endotoxin) of the cell walls of Gram-negative organisms can result in endotoxic shock. The presence of endotoxin leads to activation of various inflammatory cascades (Fig. 39.1). Release of cytokines results in vasodilation and permeability, causing a fall in blood pressure (septic shock). Activation of the clotting cascade may result in disseminated intravascular coagulopathy (DIC) with bleeding and thrombosis occurring simultaneously.

Endotoxaemia can occur in the absence of bacteria in the blood stream, for example, from Gram-negative bacteria resident in the gut, endotoxin passing across the gut wall. Gram-positive organisms, for example, pneumococci, can result in a clinical picture very similar to endotoxic shock, despite the fact that Gram-positive organisms do not contain endotoxin in their cell walls. Greater understanding of the immunological pathways leading to endotoxic shock have led to the design of monoclonal antibodies and drugs, which may block the development of endotoxic shock. Some of these products are currently being assessed in clinical practice.

AETIOLOGY

Septicaemia is almost invariably a complication of localized infection (e.g. pneumonia, meningitis). The isolation and identification of bacteria from the blood is an important aid to the diagnosis of local infection. The common causes of septicaemia and the likely organisms are shown in Tables 39.1–39.4.

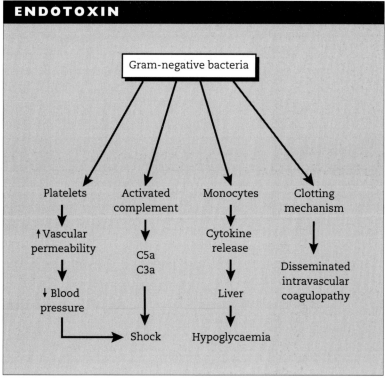

Fig. 39.1 Endotoxin activates several immune mechanisms and affects the clotting pathway.

The relative incidence of organisms causing septicaemia varies in different countries and with patients' age and circumstances, for example:
• coagulase-negative staphylococci: most frequent blood culture isolate in hospitalized patients in developed countries; related to frequent use of intravascular catheters (Plates 1, 61, 62);
• *E. coli*: frequent cause of septicaemia in elderly patients with urinary tract infections;
• *Salmonella* species: uncommon in developed countries, but an important cause of septicaemia in developing countries.

LABORATORY DIAGNOSIS

• Blood for culture should be obtained aseptically. Bacteria which are part of the normal skin flora (e.g. coagulase-negative staphylococci) may contaminate blood cultures occasionally and must be distinguished from true infection by clinical assessment.

GRAM-POSITIVE COCCI

Organism	Common focus of infection	Notes
Staphylococcus aureus	Deep abscesses, osteomyelitis, septic arthritis, vascular catheter infection, endocarditis	
Coagulase-negative staphylococci (e.g. *Staphylococcus epidermidis*)	Vascular catheter infection; other prosthetic device infections	Frequent contaminant of blood cultures, therefore clinical assessment important in confirming diagnosis
	Endocarditis	Often post-cardiac valve replacement
β-haemolytic streptococcus group A	Cellulitis, necrotizing fasciitis, puerperal sepsis, pharyngitis	Severe acute presentation; high mortality
β-haemolytic streptococcus groups C/G	Cellulitis	
β-haemolytic streptococcus group B	Septicaemia (without focus), meningitis	Important pathogen in neonatal period
Streptococcus milleri	Deep abscesses	Particularly liver, intra-abdominal, lung and brain
Streptococcus pneumoniae	Pneumonia, otitis media, meningitis	More frequent in splenectomized patients
Viridans streptococci	Endocarditis	Often subacute presentation

Table 39.1 Causes of septicaemia: Gram-positive cocci.

GRAM-POSITIVE BACILLI

Organism	Common focus of infection	Notes
Diphtheroids	Prosthetic device infection	Less common than *Staphylococcus epidermidis*
Listeria monocytogenes	Septicaemia (without focus), meningitis	Important in neonates and immunocompromised patients
Clostridium perfringens	Gas-gangrene anaerobic cellulitis	Uncommon, sometimes associated with diabetes mellitus

Table 39.2 Causes of septicaemia: Gram-positive bacilli.

GRAM-NEGATIVE COCCI

Organism	Common focus of infection	Notes
Neisseria meningitidis	Septicaemia (without focus) Meningitis	Often acute and severe with purpuric rash; endotoxic shock a common feature
Neisseria gonorrhoea	Genital infection	Septicaemia is rare (<3% of cases of gonorrhoea) associated with arthritis and skin rash

Table 39.3 Causes of septicaemia: Gram-negative cocci.

GRAM-NEGATIVE BACILLI

Organism	Common focus of infection	Notes
Escherichia coli and other coliforms	Urinary tract infection; intra-abdominal sepsis (e.g. appendicitis, cholangitis); nosocomial pneumonia	
Pseudomonas aeruginosa	Nosocomial infections	Pneumonia in ventilated patients
Haemophilus influenzae type b	Meningitis, epiglottitis	Particularly in children <5 years
Salmonella typhi and *paratyphi*	Septicaemic illness	Rare in UK, cases normally imported
Other *Salmonella* species	Enteritis	Complication of *Salmonella* enteritis
Brucella species	Septicaemic illness	Chronic illness often presenting as a pyrexia of unknown origin

Table 39.4 Causes of septicaemia: Gram-negative bacilli.

- The traditional method for detecting growth in blood cultures is by regular observation for turbidity (see Plate 42). The length of time before growth is detectable is dependent on organism type, the number of bacteria, and whether antibiotics were present in the original sample; most blood cultures become turbid within 48 h. A variety of automated blood culture instruments are now available which detect bacterial growth by various techniques, e.g. the detection of carbon dioxide by radiometric or optical methods.
- When bacterial growth is detected, Gram-staining gives a presumptive indication of the likely organism. Further identification and antibiotic susceptibility testing are carried out after the organism has been grown.

MANAGEMENT

Management of septicaemia involves supportive therapy (e.g. intravenous fluids to prevent shock) and antimicrobial agents. The choice of antibiotic is dependent on a number of factors:

- the clinical symptoms and signs suggesting the likely focus and cause of infection;

- the patient's underlying condition;
- the blood culture results, including the direct Gram stain;
- the severity of the patient's condition;
- severely ill septicaemic patients require high-dose broad-spectrum antibiotics to cover all possibilities.

Antibiotic therapy can be adjusted according to the results of culture, bacterial identification and antibiotic susceptibility testing (see Plate 45).

CHAPTER 40

Infections of the Heart

INFECTIVE ENDOCARDITIS

Definition. Infection of the endocardium of the heart, usually including valves.

Epidemiology. Infective endocarditis is uncommon (approximately 1000 cases per year in the UK). Mortality is around 20% but is increased in cases diagnosed late. In about two-thirds of patients there is a pre-existing heart defect, either congenital or acquired.

Pathogenesis. Important causes of valvular damage include rheumatic fever and congenital abnormalities; a damaged or prosthetic valve encourages deposition of fibrin, to which organisms can attach during episodes of transient bacteraemia, e.g. following operative procedures such as dental manipulation or gastrointestinal surgery.

Clinical features
- Infective endocarditis may be an acute rapidly progressive disease, or subacute with a slow indolent course, and is characterized by fever, night sweats, malaise, anaemia and anorexia.
- Circulating immune complexes result in skin lesions: splinter haemorrhages under the nails and tender nodular lesions on the palms or fingers (Osler's nodes) and glomerulonephritis (microscopic haematuria).
- Infective emboli can result in infarction of various organs, including kidney, intestine or brain. Rarely, these may be the presenting feature. Septic emboli can also cause macular lesions on the palms and soles (Janeway lesions), which consist of bacteria, necrosis and haemorrhage. Damage to the valve may result in a changing murmur, cardiac failure and valve perforation.

Causative organisms. Important causes of infective endocarditis and their source are shown in Table 40.1. Other organisms, including *Escherichia coli*, other coliforms, *Pseudomonas* species and β-haemolytic streptococci are

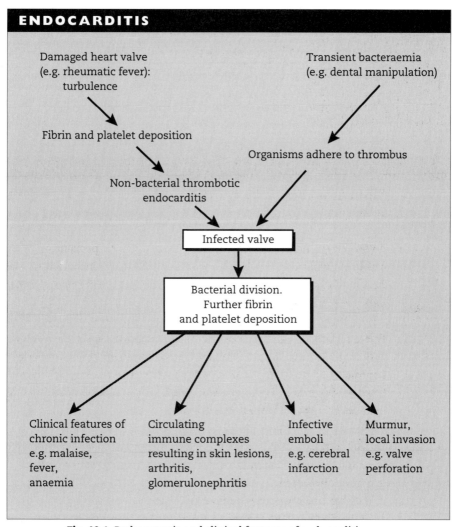

Fig. 40.1 Pathogenesis and clinical features of endocarditis.

rare causes of infective endocarditis. 'Culture-negative endocarditis' is used to describe cases of infective endocarditis diagnosed clinically in which no causative organism is isolated. This may be the result of antibiotic treatment prior to investigation, or to organisms which do not grow readily in laboratory culture media, e.g. *Coxiella burnetii* (the cause of Q-fever), *Chlamydia psittaci* (the cause of psittacosis), and *Chlamydia trachomatis*.

Laboratory diagnosis
- *Non-specific tests*: a raised white cell count and increased inflammatory

MAJOR MICROBIAL CAUSES

Organism	Source	Notes
Viridans streptococci, e.g. *Streptococcus sanguis*, *S. mitis*	Oral cavity	Responsible for approximately 60% of cases
Enterococci	Gastrointestinal tract	
Staphylococcus aureus	Skin; local infection	Associated with intravenous drug abuse; may occur on a normal valve
Staphylococcus epidermidis	Skin	Leading cause of prosthetic valve endocarditis, particularly in first few months after surgery
Candida albicans	Oral cavity; skin	Has been associated with drug abuse

Table 40.1 Major microbial causes of infective endocarditis.

markers (erythrocyte sedimentation rate (ESR) and C-reactive protein) support the diagnosis of endocarditis.
• *Cardiological investigations*: echocardiography, either transthoracic or via an electrode placed down the oesophagus (transoesophageal echocardiography) allows the visualization of vegetations on the cardiac valves.
• *Microbiological investigations*.
   (a) Blood cultures: the number of bacteria circulating in the blood is often small in endocarditis and at least three separate blood culture sets should be taken before antibiotic therapy is commenced.
   (b) Serological tests: *Chlamydia psittaci* and *Coxiella burnetii* are diagnosed by serological tests.

Management (Table 40.2). The principles of antibiotic therapy in endocarditis are:
• to use combinations of bactericidal antibiotics, which preferably show synergy for prolonged periods (4–6 weeks);
• to monitor therapy carefully by following clinical response and levels of inflammatory markers (ESR, C-reactive protein);
• to perform extended microbiological tests on the organism including minimum inhibitory concentrations, minimum bactericidal concentrations and serum back-titrations.

MANAGEMENT

Organism	Treatment
Penicillin-sensitive streptococci	Benzylpenicillin
Reduced penicillin-sensitive streptococci and enterococci	Benzylpenicillin plus gentamicin
Flucloxacillin-sensitive staphylococci	Flucloxacillin plus gentamicin or flucloxacillin plus fucidin
Flucloxacillin-resistant staphylococci	Vancomycin plus gentamicin
Fungi	Amphotericin
Chlamydia psittaci/Coxiella burnetii	Tetracycline

Table 40.2 Antimicrobial treatment of infective endocarditis.

Management of cardiac aspects of the condition may involve treatment for cardiac failure and in some cases, surgery to replace the infected, damaged valve.

Prevention. Prophylactic antibiotics are used for high-risk groups undergoing operative procedures, including dental operations.

MYOCARDITIS

Definition. Infection of the myocardium.

Aetiology. Coxsackie B viruses are the most common causes of viral myocarditis; others include cytomegalovirus, Epstein–Barr virus and mumps virus. Bacteria rarely cause myocarditis.

Clinical features. Fever, chest pain, dyspnoea; chest X-ray may show an enlarged heart; a pericardial effusion is often present; electrocardiographic (ECG) changes are a frequent feature.

Laboratory diagnosis. This is by:
- viral isolation from pharyngeal washings or faeces;
- myocardial biopsy may show typical histological features and occasionally viruses may be grown from the tissue;
- serology (a fourfold rise in antibody titres or evidence of an IgM response).

Treatment. Antiviral therapy is available for some causes, e.g. ganciclovir for cytomegalovirus (CMV) infection.

PERICARDITIS

Definition. Inflammation of the pericardium resulting from infection.

Aetiology
- *Viral*: enteroviruses (Coxsackieviruses A and B, Echoviruses), CMV, Epstein–Barr virus; occasionally, other viruses.
- *Bacteria*: S. aureus, S. pneumoniae, β-haemolytic streptococci, Mycobacterium tuberculosis. Bacterial pericarditis may result from haematogenous spread from a distant primary site of infection or, less commonly, direct extension from a primary lung infection.
- *Fungal*: *Histoplasma*; *Aspergillus* and *Candida*, as part of invasive fungal disease in the immunocompromised.

Clinical features. These include fever, chest pain, dyspnoea; a pericardial rub (before the pericardial effusion develops); and signs of cardiac failure. Chest X-ray may show an enlarged heart with a pericardial effusion. Patients with bacterial pericarditis normally have an acute presentation and are generally more toxic than patients with viral pericarditis.

Laboratory diagnosis
- Viral: isolation from pharyngeal washings, faeces or pericardial fluid; serology.
- Bacterial: culture of pericardial fluid and blood cultures.

Management
- *Viral*: specific antiviral therapy, e.g. ganciclovir for CMV disease; pericardial drainage may be necessary.
- *Bacterial*: pericardial drainage and occasionally pericardectomy together with appropriate antibiotics, depending upon pathogen.
- *Tuberculous*: drainage and pericardectomy; antituberculous drugs plus corticosteroids to reduce inflammation.
- Fungal: amphotericin; pericardial drainage may be necessary.

CHAPTER 41

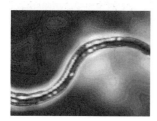

Perinatal and Congenital Infections

Definitions. Although congenital and neonatal infections are often described separately, there is considerable overlap in aetiology and pathogenesis.

Congenital infections: these are infections acquired *in utero*, either across the placenta or from the genital tract (ascending infection). Infections acquired in the first trimester are normally acquired transplacentally and cause developmental defects (congenital malformations); these may result in spontaneous abortion, or a neonate born with mild or severe defects. Infections acquired later in pregnancy may result in stillbirth or a neonate with an active infection.

Neonatal infections: infection during the neonatal period may be acquired *in utero* or from the genital tract during delivery and typically presents early in the neonatal period. Late-onset neonatal infections may be acquired from attending adults (e.g. parents, nursing staff), other children or the hospital environment.

CONGENITAL INFECTIONS (Table 41.1)

Diagnosis. This may be difficult with some intrauterine infections as maternal infection may be asymptomatic and clinical manifestations in the infant may only appear months or years following birth. The important microbiological investigations are:
* confirmation of infection in the mother during pregnancy, principally by serological tests;
* isolation of the organism from the fetus or placenta, or from samples taken from the neonate (e.g. blood, faeces, urine, cerebrospinal fluid (CSF));
* serological tests on neonatal blood.

The presence of raised levels of specific IgG class antibodies may reflect merely the presence of maternal antibodies, as IgG antibodies can cross the placenta. However, the presence of specific IgM class antibodies, is diagnostic of intrauterine acquired infection, as maternal IgM antibodies cannot cross the placenta.

IMPORTANT CONGENITAL CONDITIONS

Organism	Congenital malformation
Rubella virus	Congenital heart defects; cataracts; microcephaly; deafness; hepatosplenomegaly, purpura
Cytomegalovirus	Deafness; microcephaly; low birth weight
Varicella-zoster virus	Low birth weight; encephalitis, limb hypoplasia; neonatal chickenpox (severe and widespread)
Human immunodeficiency virus	Paediatric AIDS
Treponema pallidum	Hepatosplenomegaly; bone and tooth defects; microcephaly
Toxoplasma gondii	Choroidoretinitis; hydrocephalus; microcephaly
Listeria monocytogenes	Intrauterine death; neonatal infection (meningitis)

Table 41.1 Important congenital conditions.

Rubella virus

Epidemiology. The risk of developing congenital infection following maternal rubella virus infection depends on the stage of pregnancy in which the maternal infection occurs; multiple severe defects are most common when infection occurs during the first 8 weeks of pregnancy. Since the introduction of widespread or universal rubella vaccination, the incidence of congenital rubella syndrome has decreased.

Clinical features. Rubella affects blood vessels in developing organs, causing: congenital heart defects; eye defects (cataracts, retinopathy, microophthalmos); deafness; microcephaly, seizures and mental retardation; hepatosplenomegaly; and thrombocytopaenia.

Laboratory diagnosis. By: the presence of rubella-specific IgM antibodies in the mother during early pregnancy; isolation of virus from the urine or throat of an infected neonate; and the presence of rubella-specific IgM antibodies in cord blood.

Prevention. Vaccination of girls and identification of non-immune women by antenatal screening followed by vaccination postpartum (the vaccine contains live virus and therefore cannot be given during pregnancy) has led to a significant reduction in both the number of rubella-susceptible women, and congenital rubella in developed countries. In the USA and more recently also in the UK, boys are also vaccinated in an attempt to lower further the pool of rubella-susceptible individuals in the community.

Cytomegalovirus (CMV)

Epidemiology
- Given the successful vaccination strategy against rubella virus infection, congenital infection caused by CMV is now the most common intrauterine infection (<1% of neonates in the UK; 1–2% in the USA).
- CMV intrauterine infection may follow primary CMV infection, or less commonly, reactivation of latent infection (secondary infection). Fetal damage may occur following maternal infection in any trimester, but is more serious when infection occurs early in pregnancy; about 40% of fetuses become infected following primary infection during pregnancy, *but* 90% will be normal at birth and only a small percentage of these will later present with deafness and/or mental retardation.

Clinical features. These include: pneumonitis; hepatosplenomegaly; thrombocytopaenia; anaemia; low birth weight; microcephaly; cerebral palsy; deafness; and choroidoretinitis.

Laboratory diagnosis
- Maternal CMV infection in pregnancy is diagnosed by detection of CMV-specific IgM.
- Neonatal infection is confirmed by the detection of CMV-specific IgM and the isolation of the virus from throat swabs or urine. Excretion of CMV in the urine may continue for several months, and babies with congenital CMV infection should be isolated whilst in hospital.

Prevention. The strategy followed for congenital rubella prevention is not possible as there is currently no vaccine available for CMV. For this reason, and as severe disease after intrauterine infection is rare, antenatal screening is not performed routinely.

Toxoplasmosis

Epidemiology. Toxoplasma gondii is a protozoon whose definitive host is the cat. Humans become infected via ingestion of oocysts from contact with cat faeces, or from ingestion of meat contaminated with trophozoites.

Maternal infection is often subclinical or presents as a mild influenza-like illness. In the UK toxoplasmosis affects about 0.5% of pregnancies. Fetal damage most frequently occurs in the first trimester.

Clinical features. Stillbirth; choroidoretinitis; microcephaly; convulsions; intracerebral calcification with resultant hydrocephalus; hepatosplenomegaly and thrombocytopaenia.

Laboratory diagnosis. Detection of toxoplasma-specific IgM antibodies in the mother during pregnancy, or in the neonate.

Treatment and prevention. Spiramycin or pyrimethamine may reduce the severity of sequelae. There is no vaccine for toxoplasma, but pregnant women are advised to avoid contact with cat faeces during pregnancy. Antenatal screening for toxoplasma antibodies is performed in some countries.

Varicella-zoster virus

Maternal chickenpox around the time of delivery may result in severe neonatal infection; in such circumstances, neonates should be given specific immunoglobulin to varicella-zoster (ZIG) and/or acyclovir immediately after birth. Rarely, maternal infection in the first trimester may result in congenital defects (low birth weight, encephalitis, choroidoretinitis, hyperplastic limbs). Non-immune pregnant women should be advised to avoid contact with cases of chickenpox; in cases of proven contact and sero-negativity of pregnant women, they should be treated with ZIG.

Parvovirus

Maternal parvovirus infection results in spontaneous abortion in about 10% of cases. The virus infects erythrocyte precursors, resulting in anaemia and cardiac failure in the fetus; congenital malformations have not been reported. Maternal infection with parvovirus can be confirmed by the detection of specific IgM. There is no vaccine to parvovirus. Pregnant women should be advised to avoid contact with children with suspected parvovirus infection.

Syphilis

Epidemiology. Congenital syphilis is now rare in developed countries because of routine antenatal screening, but remains a serious problem in developing countries. Primary or secondary syphilis occurring during pregnancy may result in intrauterine infection.

Clinical features. Stillbirth; hepatosplenomegaly; lymphadenopathy; skin rashes; nasal discharge; bone, tooth and cartilage defects (e.g. saddle-shaped nose); and deafness.

Laboratory diagnosis. By detection of specific IgM to *Treponema pallidum* in neonatal serum.

Prevention. There is routine antenatal screening for the presence of active syphilis in pregnant women; detection should be followed by prompt treatment with penicillin.

Tuberculosis

Intrauterine infection with tuberculosis remains a problem in areas of the world where the disease is common. Infection early in pregnancy may result in abortion; later in pregnancy it may result in the birth of a neonate with active tuberculosis, often presenting as pneumonia or hepatitis.

Listeria monocytogenes

Listeria monocytogenes is a commensal of the gastrointestinal tract. Maternal infection (asymptomatic or influenza-like illness) may result in intrauterine infection, with abortion or stillbirth. Infection late in pregnancy may result in the birth of an infected neonate with septicaemia, and meningitis.

Malaria

Congenital or neonatal malaria occurs most frequently when the mother becomes infected for the first time during pregnancy. Infection may result in abortion or stillbirth, or premature delivery. Neonates may not develop symptoms of malaria for several weeks following delivery.

NEONATAL INFECTIONS

Neonates, particularly when premature, are immunocompromised because of an immature immune system. Minor sepsis can lead rapidly to septicaemia and shock, with a high mortality. Advances in intensive care have led to increased numbers of premature neonates surviving. This has increased the use of long-term intubation, indwelling intravascular lines and other monitoring devices, and antibiotic treatment. Early diagnosis of infection is difficult; signs of sepsis may be absent or obscure, e.g. decreased alertness, temperature instability, reduced feeding.

Group B β-haemolytic streptococci

These are commensals of the adult genital tract and perianal area; about 30–50% of pregnant women colonized. Subsequent colonization of newborn babies is common, but only a small number become infected (<1 : 1000); premature and low birth weight infants are particularly at risk. Infections may present early (less than 48 h post-delivery) or late (1 week to 6 months post-delivery). Typically, early-onset infections are more severe, with pneumonia and septicaemia; late-onset infections often present as meningitis. Treatment is penicillin with gentamicin.

Listeria monocytogenes

This is acquired perinatally, probably from the maternal intestinal tract. It is an important cause of neonatal meningitis and septicaemia during the first few weeks of life and also a cause of septic abortion. Treatment is with ampicillin or penicillin.

Escherichia coli

E. coli is an important cause of neonatal meningitis and septicaemia; acquisition is probably from the maternal intestinal tract. Invasive infections are associated with certain *E. coli* serotypes, particularly the K1 serotype. Treatment is various, according to antibiotic susceptibility tests, e.g. cefuroxime or gentamicin.

Other Gram-negative aerobic bacilli

Klebsiella, Pseudomonas aeruginosa and other antibiotic-resistant Gram-negative bacilli can be acquired from the hospital environment and are often associated with prolonged hospitalization in special-care baby units. Infections include pneumonia, septicaemia, meningitis and urinary tract infections. Treatment is often with antibiotic combinations, guided by antibiotic sensitivity tests.

Coagulase-negative staphylococci

The increased use of indwelling intravascular catheters during the neonatal period has led to an increase in the number of associated infections caused by coagulase-negative staphylococci, particularly *Staphylococcus epidermidis*. *S. epidermidis* is the most common organism isolated from neonatal blood cultures. Management includes possible removal of infected catheters and the use of antibiotics such as vancomycin or teicoplanin.

Staphylococcus aureus

This organism may cause pneumonia, intravascular catheter infections, skin infections (e.g. scalded skin syndrome) and conjunctivitis. Treatment is usually with flucloxacillin.

Neonatal eye infections

Aetiology and epidemiology. Neisseria gonorrhoea and *Chlamydia trachomatis* are acquired from the maternal genital tract; maternal infection may be subclinical. Other causes include *S. aureus*, *Haemophilus influenzae*, pneumococci which are acquired from the infant's own flora or from close contacts.

Clinical features
- *N. gonorrhoea*: severe conjunctivitis with pus and peri-orbital inflammation; frequently presents within 48 h of birth.
- *C. trachomatis*: less severe than gonococcal conjunctivitis; often presents 3–7 days after delivery; if untreated, chlamydial pneumonitis may occur as a late complication (1–3 months).
- *S. aureus*, *H. influenzae*, *S. pneumoniae*; purulent conjunctivitis; normally present >1 week post-delivery.

Laboratory diagnosis
- *N. gonorrhoea*: microscopy (intracellular Gram-negative diplococci) and culture of conjunctival swab; appropriate investigations for maternal infection, including cervical swabs.
- *Chlamydia*: culture of conjunctival swab or direct detection of antigen by fluorescence or enzyme-linked immunosorbent assay (ELISA); appropriate investigations for maternal infection, including cervical swabs.

- *S. aureus, H. influenzae, S. pneumoniae*: conjunctival swabs for routine bacterial culture.

Treatment and prevention
- *N. gonorrhoea*: penicillin, spectinomycin or cefotaxime; prevention is by antenatal screening and treatment of pregnant women with gonorrhoea. The routine administration of parenteral penicillin or antibiotic eyedrops (chloramphenicol) to all newborn babies is carried out in areas of the world with a high incidence of gonorrhoea.
- *C. trachomatis*: tetracycline eye ointment plus oral erythromycin.
- *S. aureus, H. influenzae, S. pneumoniae*: antibiotics (e.g. chloramphenicol), eyedrops or ointment; systemic antibiotics are sometimes required (e.g. flucloxacillin for *S. aureus*).

Herpes simplex virus

Epidemiology. Acquired from the maternal genital tract during delivery; asymptomatic shedding of herpes simplex virus occurs in about 0.1% of pregnant women; the incidence is higher in women of lower socio-economic groups. Maternal infection may be primary or recurrent. The risk of acquisition is increased in maternal primary infections because excretion of virus is higher than in recurrent infection.

Clinical features. Encephalitis or meningoencephalitis; generalized infection with hepatitis; and conjunctivitis.

Laboratory diagnosis. This is by: direct immunofluorescence of vesicular fluid from skin lesions; culture of virus from throat washings, CSF or conjunctival swabs.

Treatment and prevention. Treatment of herpes encephalitis with high doses of acyclovir has reduced mortality to around 25%. Clinical genital herpes at term is an indication for caesarian section.

Preventative strategies based on routine antenatal screening for genital herpes are impractical because virus shedding is intermittent and the cost would be high; screening of high-risk groups has been advocated in some countries.

Human immunodeficiency virus (HIV)

Epidemiology. Most cases of paediatric acquired immune deficiency syndrome (AIDS) are a result of vertical transmission from an infected mother. The baby may acquire the virus transplacentally, during delivery, or, less commonly, via breast feeding. Risk of transmission has been estimated to be 15–40%.

Clinical features. Poor weight gain, lymphadenopathy, hepatosplenomegaly, and an increased susceptibility to infection.

Laboratory diagnosis. This is dependent on the detection of HIV antibodies, but these may be maternally acquired. A definitive diagnosis can only be confirmed if antibodies are still present after 1 year. The development of HIV-specific IgM serological tests, the detection of HIV antigenaemia and the detection of HIV RNA or proviral DNA by gene amplification techniques (polymerase chain reaction) has significantly improved early diagnosis. Treatment of HIV-infected pregnant women with AZT significantly reduces vertical HIV transmission.

Hepatitis B virus (HBV)

Epidemiology. Vertical transmission of HBV is an important route of infection. Worldwide vertical transmission is the most frequent way. Acquisition rates vary according to hepatitis B e-antigen status. In many developed countries transmission rates are <10%, but acquisition rates rise to around 90% in South-East Asia. Transmission occurs mostly at the time of delivery.

Clinical features. Infection is often subclinical, but the infant becomes a chronic carrier with subsequent cirrhosis and a high risk of developing liver cancer in later life.

Laboratory diagnosis. This is dependent on detection of hepatitis B surface antigen in the infant's blood, which persists for >6 months.

Prevention. Administration of specific HBV immunoglobulin as early as possible after delivery and simultaneous active immunization with hepatitis B vaccine (given at birth, and 1 and 2 months after birth).

OBSTETRIC INFECTIONS

Puerperal sepsis

This is a uterine infection following delivery or termination of pregnancy, and is now uncommon in developed countries. Important pathogens include *S. pyogenes*, *C. perfringens*, *E. coli* and *Bacteroides* spp. Predisposing factors include: instrumentation; premature rupture of membranes; retained placental material, and the use of non-sterile equipment/poor practice, e.g. illegal abortion. Laboratory investigations include culture of high vaginal swab and blood cultures. Treatment is with combinations of antibiotics to cover likely pathogens (e.g. penicillin, gentamicin plus metronidazole).

CHAPTER 42

Miscellaneous Viral Infections

MUMPS

Aetiology and epidemiology. Mumps virus is a paramyxovirus. Transmission is by droplet spread or contact with saliva. It has worldwide distribution.

Clinical features. The incubation period is 18–21 days. There is a short prodromal phase (48 h) of malaise and fever, followed by parotid gland enlargement. Infectivity lasts from three days before to three days after symptoms subside.

Complications
- Meningitis, often mild in <5% of cases.
- Epididymo-orchitis in about 25% of adult males infected with the virus. It is an occasional cause of sterility.
- Pancreatitis, myocarditis, arthritis (rare), oophoritis.

Laboratory diagnosis. By virus isolation from saliva or cerebrospinal fluid (CSF) and serology: detection of mumps-specific IgM antibody.

Prevention. Attenuated live virus vaccine is normally given in combination with measles and rubella vaccines (MMR) in childhood immunization schemes.

MEASLES

Aetiology and epidemiology. Measles virus is a paramyxovirus with a worldwide distribution. Transmission is by droplet spread. It is highly infectious and the transmission rate to non-immune contacts is high.

Clinical features. The incubation period is 9–10 days. There is a prodromal phase (48 h) of fever, cough, rhinitis and conjunctivitis, followed by the appearance of a maculopapular rash starting on the face and trunk, and spreading peripherally; lesions also present on the mucosa, particularly in the mouth (Koplik spots).

Complications
- Pneumonia, particularly in adults and malnourished children.
- Encephalitis and subacute sclerosing panencephalitis.
- Otitis media (bacterial superinfection, e.g. Staphylococci, Streptococci).

Laboratory diagnosis. This is by virus isolation from nasopharyngeal secretions and serology.

Prevention. Attenuated live virus vaccine is normally given in combination with rubella and mumps vaccines (MMR).

CHICKENPOX AND SHINGLES

Aetiology. Chickenpox and shingles are caused by varicella-zoster virus, a member of the herpesvirus family. Primary infection with varicella-zoster virus results in chickenpox; the virus then persists in dorsal route ganglia and may reactivate later in life to cause shingles (zoster).

Chickenpox

Epidemiology. Transmission is by droplet spread from virus-infected saliva or by direct contact with skin vesicles. It has a worldwide distribution.

Clinical features. The incubation period is 14 days (range 11–22 days). There is a short prodromal phase (48 h) of fever and general malaise, followed by the appearance of vesicles, first on the face and trunk, and then peripherally. Vesicles become pustules and form scabs.

Complications
- Secondary bacterial infection of vesicles, normally staphylococcal.
- Pneumonia occurs in 10–20% of adults with chickenpox.
- Meningitis and encephalitis are rare complications.

Shingles

- Reactivation of the latent virus in the dorsal route ganglia can result in crops of vesicles associated with various dermatomes, e.g. thorax or face

(Plate 63), including ophthalmic zoster; pain normally precedes the vesicular rash by several days.
- Predisposition to zoster includes increasing age and immunocompromised patients (e.g. haematological malignancy, AIDS, transplantation). In immunocompromised patients the local eruptions may spread to cause a generalized vesicular rash, with lesions in various deep tissues.
- Early treatment with acyclovir may reduce symptoms.

Laboratory diagnosis. By: examination of vesicular fluid by electronmicroscopy; virus culture; and serology (chickenpox only).

Treatment and prevention. Severe varicella-zoster infections are treated with acyclovir. Varicella-zoster-specific immunoglobulin can be given to immunocompromised contact patients. A live attenuated vaccine has recently been developed, but is not yet in routine use.

Herpes simplex virus (HSV)

Aetiology and epidemiology. HSV is a member of the herpes virus group; there are two serotypes, HSV 1 and HSV 2. They have a worldwide distribution. Transmission is via virus-infected saliva (HSV 1), sexual contact (HSV 2) or direct contact with vesicles. Following primary infection (often asymptomatic), HSV remains latent in the sensory ganglia; reactivation results in secondary infection.

Clinical features
- Stomatitis: seen in children and immunocompromised patients (see p. 219).
- Herpes labialis (cold sore): crops of vesicles around lips; results from reactivation of HSV in response to various stimuli (e.g. stress, sunlight).
- Herpetic whitlow: results from inoculation of HSV into finger (typically seen in health care workers).
- Central nervous system (CNS) infections: meningitis, encephalitis (see p. 266).
- Genital infections (see p. 259).
- Eye infections (see p. 210).
- Multisystem infections in the immunocompromised.

Laboratory diagnosis. By: viral culture of appropriate specimens; direct immunofluorescence or enzyme-linked immunosorbent assay (ELISA) for herpes antigens; and serology (of doubtful value as HSV antibodies are present in many patients).

Treatment. Acyclovir is useful for active infection, but does not eliminate latent HSV. Topical, oral and intravenous preparations may be prescribed, depending on the site and severity of infection. Prophylactic acyclovir is given to immunocompromised patients.

RUBELLA

Aetiology and epidemiology. Rubella is caused by an RNA virus. It has a worldwide distribution with transmission by respiratory droplet spread.

Clinical features. The incubation period is 14–21 days, with a prodromal phase of fever and general malaise, followed by the appearance of a maculopapular rash lasting from 3–5 days. Enlargement of the suboccipital lymph nodes is a typical feature. Mild arthralgia or arthritis may occur. Subclinical infection is common.

Complications. Congenital rubella resulting in fetal damage may occur following maternal infection during the first trimester (see p. 280).

Laboratory diagnosis. By serology for rubella-specific IgM antibodies and rarely by virus culture from throat swabs.

Prevention. A live attenuated rubella vaccine is available and is given in combination with measles and mumps vaccine (MMR) as part of childhood vaccination programmes.

CYTOMEGALOVIRUS (CMV) INFECTIONS

Aetiology and epidemiology. CMV is a member of the herpesvirus family. Infection occurs worldwide. The virus is found in saliva and transmission is related to close contact; urinary excretion of the virus in small children may be a source of infection.

Following primary infection, CMV remains latent and may reactivate during periods of immunosuppression (AIDS, organ transplantation) to cause multisystem infection. It may be transmitted via blood products or transplanted tissues.

Clinical features. Infection is frequently asymptomatic. Glandular fever syndrome may occur in acute infection of teenagers (see Epstein–Barr virus infection, below).

Complications
- Infection during pregnancy may result in fetal damage (see p. 283).
- Severe multisystem infections may occur primarily in patients who are immunodeficient (e.g. AIDS patients, transplant recipients, patients with haematological malignancies). Infections include pneumonia, retinitis (particularly in AIDS patients), hepatitis and encephalitis.

Laboratory diagnosis
- Viral culture of throat washings or urine, or from peripheral blood leucocytes in invasive infections. Tissue culture takes about 6 weeks before the presence of CMV can be recognized, but detection of viral early antigens in tissue culture cells by specific fluorescent antibody staining techniques speeds up the diagnosis to within 1–2 days (DEAFF test).
- Serology for CMV-specific IgM.
- Detection of CMV DNA by the polymerase chain reaction.

Treatment and prevention. Treatment is with ganciclovir. Prevention is important in the immunocompromised, but difficult, as latent infection is common. Strategies include screening of blood and organ donors for CMV, and the use of prophylactic ganciclovir or related compounds.

EPSTEIN–BARR VIRUS (EBV) INFECTIONS

Aetiology and epidemiology. EBV is a member of the herpesvirus family, with a worldwide distribution. Transmission is by contact with saliva. Infection is most common in young adults.

Clinical features. Fever, sore throat and local lymphadenopathy may occur with general malaise and lethargy which may continue for several months (glandular fever syndrome).

Complications
- Mild hepatitis, encephalitis (rare), autoimmune haemolytic anaemia (rare).
- EBV is associated with a number of malignancies, e.g. Burkitt's lymphoma, B-cell lymphomas in immunodeficient patients, and nasopharyngeal carcinoma.

Laboratory diagnosis
- The appearance of atypical lymphocytes in blood films.
- Serology: detection of cross-reacting antibodies to horse erythrocytes (monospot test); specific serological tests are available for the detection of EBV IgM antibody.
- EBV DNA can be detected by the polymerase chain reaction (PCR).

Treatment and prevention. No effective treatment or vaccine is available.

HUMAN HERPES VIRUS 6 INFECTION

Aetiology and epidemiology. Human herpes virus 6 (HHV 6) most frequently infects young children. Transmission is via contact with infected saliva.

Clinical features. HHV 6 causes exanthem subitum, also known as roseola infantum. After a 2-week incubation period, there is fever and a characteristic maculopapular rash. HHV 6 reactivation has been seen in immunocompromised patients.

Diagnosis and treatment. Diagnosis is by culture; serological tests are not yet widely available. The infection is mild and self-limiting; treatment and prevention are unnecessary.

CHAPTER 43

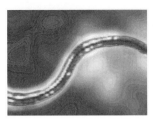

Infections in the Compromised Host

The immune system consists of adaptive (cellular and humoral) and innate mechanical barriers—skin and mucous membranes, complement and phagocytic cell mechanisms. When any of these protective mechanisms are reduced the host is immunocompromised. Immunocompromised patients are more liable to have severe infections and can become infected with either a conventional pathogen (which can also cause disease in a non-compromised individual) or opportunistic pathogens (organisms which are usually unable to cause a disease in a healthy person). Various types of defect in the immune defence mechanisms can predispose to infections with certain pathogens.

INFECTIONS ASSOCIATED WITH DEFECTS IN INNATE IMMUNE SYSTEMS (Table 43.1)

Infections associated with prosthetic devices

- *Intravenous and peritoneal dialysis catheters*: these catheters breach the skin barrier, allowing the commensal organisms access to the deeper sites. The infections are commonly associated with coagulase-negative staphylococci, S. aureus, and, less frequently, Gram-negative aerobic bacilli and Candida.
- *Prosthetic valves and joints*: Coagulase-negative staphylococci are the most common pathogens, gaining access during surgery. The organism can attach to the prosthesis, which acts as a nidus of infection.

Infections associated with burns

Burns can destroy extensive areas of the mechanical barriers of the body and also result in abnormalities in localized neutrophil function and antibody response. The burn exudates produce nutrition for micro-organisms to colonize and cause infection. The usual pathogens are: aerobic Gram-negative bacilli, Pseudomonas, including P. aeruginosa, S. aureus, Streptococcus pyogenes.

INFECTIONS

Defect/condition	Common pathogen
Intravascular and peritoneal dialysis catheters	*Staphylococcus epidermidis*, *S. aureus* (most common), coliforms; *Candida*
Urinary catheter	Coliforms
Prosthetic valves and joints	*S. epidermidis*
Burns	Coliforms, *Pseudomonas*, *S. aureus*, *Streptococcus pyogenes*
Biliary obstruction	Coliforms, *Enterococcus*
Trauma/surgery	*S. aureus*

Table 43.1 Infections associated with defects in mechanical barriers.

Infections associated with deficiencies in normal clearance mechanisms

- *Urinary catheters* allow organisms access to the bladder, overcoming defences, including urine washout. Bacteria can grow up the inside of a urinary catheter. The infections are commonly associated with Gram-negative aerobic bacilli from the patient's own faecal or periurethral flora.
- The presence of stones in the kidney, common bile and salivary ducts can result in infections proximal to the obstruction.
- In the respiratory tract there may be damage to the cilia, e.g. with cystic fibrosis patients, which predisposes to pneumonia with organisms such as *S. aureus*, *H. influenzae* and *P. aeruginosa*.

Infections associated with defects in complement and phagocytic activity

These defects are normally congenital. Defects in phagocytic function may affect chemotaxis, phagocytosis or bacterial killing; associated infections often involve pyogenic bacteria, e.g. *S. aureus*. Defects in the complement system are associated with bacterial infections, e.g. patients with defects in the later complement components (C7–9) have an increased risk of meningococcal and gonococcal infections.

INFECTIONS RELATED TO CELLULAR AND HUMORAL IMMUNODEFICIENCY

The type of underlying immunodeficiency (Table 43.2) may be related to the nature and severity of any subsequent infection (Table 43.3). Increasingly, these infections are iatrogenic and are caused by opportunistic pathogens. The length of time a patient is immunosuppressed may influence the type of associated infection.

FACTORS AFFECTING ADAPTIVE IMMUNITY

Type	Defect (example)	Notes
Primary	B- and T-cell combined immunodeficiency	
Secondary	Chemotherapy	e.g. Cyclophosphamide
	Infections	e.g. HIV-1 and -2
	Malignancy	e.g. Leukaemia, lymphoma
	Splenectomy	Reduced levels of immunoglobulin subclasses
	Irradiation	Affects proliferation of lymphoid cells
	Corticosteroids	

Table 43.2 Factors affecting adaptive systems.

INFECTIONS

Defensive defect	Examples of underlying condition	Opportunistic organism
T-cell	Lymphoma	Varicella-zoster virus
		Candida
		Toxoplasma gondii
T- and B-cell	Chronic lymphatic leukaemia	Varicella-zoster virus
	Immunosuppressive drugs	Candida
		Toxoplasma gondii
		Cytomegalovirus
		Pneumocystis carinii
Neutropaenia	Acute leukaemia	Staphylococcus aureus
	Cytotoxic treatment	Aspergillus fumigatus
		Coliforms
		Cytomegalovirus
		Pneumocystis carinii

Table 43.3 Examples of infections related to deficiency in cellular and humoral immunodeficiency.

Table 43.4 Opportunistic pathogens associated with solid organ transplant patients.

OPPORTUNISTIC PATHOGENS

Bacteria
Staphylococcus aureus
Coagulase-negative staphylococci
Mycobacterium avium-intracellulare
Enterobacteriaceae

Fungi
Candida species
Aspergillus species
Pneumocystis carinii

Protozoa
Toxoplasma

Parasites
Strongyloides stercoralis

Viruses
Herpes simplex viruses
cytomegalovirus, varicella-zoster virus, Epstein–Barr virus

Infections in some immunocompromised patients can be multifactorial and associated with several factors, e.g., in recipients of solid organ transplants, the type of organ transplant; the immune suppression required to prevent rejection of a grafted organ; the pathogens to which a recipient is exposed, including from donor organ and the environment. Some infections associated with organ transplant patients are shown in Table 43.4.

CHAPTER 44

Zoonoses

Zoonoses are human infections acquired from animals. The causative agents are transmitted by direct contact via insects (inhalation), or by the faecal–oral route. Domesticated animals are predominant reservoirs of disease in Europe; wild animals in tropical countries. Many source animals may not have an obvious illness, but may still be infective to man. Zoonoses may be classified according to the mechanism of transmission (Table 44.1). Many parasitic infections are zoonoses; these are described in Chapters 21 and 22.

BRUCELLOSIS

Epidemiology. Causative organisms are *Brucella abortus* (cattle), *B. melitensis* (goats), and *B. suis* (pigs), which have a worldwide distribution. *B. abortus* causes abortion in cows. Humans become infected following consumption of unpasteurized milk; veterinary surgeons, farmers, abattoir workers from direct contact with urine, vaginal secretions or products of conception derived from infected cows. *B. melitensis* is acquired from unpasteurized milk or dairy products from infected goats and sheep, particularly in Mediterranean countries. *B. suis* infections (mainly in USA) are associated with the handling of infected pig carcasses.

Clinical features. The incubation period is variable, usually 1–2 months. Brucellosis is an important cause of pyrexia of unknown origin. Bacteria pass to lymph nodes, cause septicaemia and infect the liver, spleen and bone marrow. Acute presentation includes undulant fever, sweats, arthralgia, myalgia. Chronic presentation involves periods of fever, splenomegaly, depression, weight loss.

Laboratory diagnosis. By blood cultures and serology.

Treatment and prevention. Tetracycline, streptomycin or co-trimoxazole are prescribed for up to 3 months. Preventative measures include by pasteurization of milk; reducing the animal reservoir by immunization and

HUMAN INFECTIONS FROM VERTEBRATES

Method of Transmission	Pathogen	Example	Infection
Direct contact	Brucella	During calving from infected cow	Brucellosis
	Bacillus anthracis	Contact with anthrax spores on contaminated hides	Anthrax
Food	Salmonella	Food and dairy products	Enteritis
	Campylobacter	Food and dairy products	Enteritis
	Brucella	Dairy products	Brucellosis
	Coxiella burnetii	Dairy products	Q-fever
Faecal–oral	Cat and dog faeces	Toxocara cati and canis	Toxocariasis
	Cat faeces	Toxoplasma gondii	Toxoplasmosis
Saliva: bite	Rabid animal	Rabies virus	Rabies
	Rat	Streptobacillus moniliformis	Rat bite fever
	Dog	Pasteurella multocida	Cellulitis
Urine into conjunctiva or other mucosal surface	Rat urine-contaminated water	Leptospira	Leptospirosis
	Bush rat urine	Arenavirus	Lassa fever

Table 44.1 Human infections from vertebrates.

selective slaughtering; wearing of protective clothing in high-risk situations.

LEPTOSPIROSIS

Epidemiology and pathogenesis. This is caused by *Leptospira interrogans* which infects animals worldwide, such as rats. Approximately 50 cases occur in the UK and 100 cases in the USA per annum. Humans are infected from contact with contaminated urine of various animals: rodents and occasionally farm animals, including pigs, cattle and dogs. Spirochaetes may penetrate abraded or intact skin, conjunctival or mucous membranes, or intestinal mucosa following ingestion, and localize at various sites. At-risk groups include sewage workers, farmers and water-sport enthusiasts.

Clinical features. The incubation period is 5–14 days. An influenza-like illness which is often self-limiting but may progress and result in hepatitis, uraemia or meningitis.

Laboratory diagnosis. Blood or urine samples may be examined by dark ground microscopy and cultured; serology is the principal method of diagnosis.

Treatment and prevention. Benzylpenicillin or tetracycline. Preventative measures include the covering of cuts and abrasions in at-risk groups, and rodent control.

ANTHRAX

Epidemiology. This infection is caused by *Bacillus anthracis*, a pathogen primarily of sheep and cattle; human infection is rare in developed countries. Bacilli are excreted in faeces, urine and saliva of infected animals, and spores may survive in contaminated soil for many years. It is an occupational disease of workers handling sheep wool, bone meal or cattle hides which may be contaminated with spores. Spores enter the body via skin or mucous membranes, or are occasionally inhaled.

Clinical features. Cutaneous anthrax results in the formation of a black necrotic ulcer, lymphadenopathy and, occasionally, septicaemia. Pulmonary anthrax is rare and follows inhalation of spores.

Laboratory diagnosis. By microscopy and culture of skin lesions; sputum microscopy and culture.

Treatment and prevention. Treatment is with high-dose penicillin. Prevention includes decontamination of imported hides and wool; cremation of carcasses of infected animals; and vaccination of at-risk workers and animals.

PLAGUE

Epidemiology. Infection caused by *Yersinia pestis*, with worldwide distribution, but which is now rare in UK and uncommon in USA. Rodents, particularly rats, are the animal reservoir. The infection is spread by the rat flea, *Xenopsylla cheopsis*, from rat-to-rat or rat-to-human.

Pathogenesis
- *Bubonic plague*: *Y. pestis* multiplies in the flea gut. When the flea is feeding

on a host, infected material is regurgitated into the bite wound. Bacterial multiplication occurs at the entry site and spreads to lymphatics. Lymph nodes enlarge and form 'buboes'.

- *Pneumonic plague*: droplet spread from person to person.

Clinical features. The incubation period is 2–6 days following, fever, lymphadenopathy, occasionally, septicaemia and multi-organ failure. Pneumonic plague causes severe pneumonia with septicaemia, septic shock and haemorrhage. If untreated, mortality is 50% for bubonic plague and 90% with pneumonic plague.

Laboratory diagnosis. By microscopy and culture of lymph node aspirates or sputum blood cultures.

Treatment and prevention. Treatment is with streptomycin and tetracycline. Prevention is by rodent control, strict isolation of patients with pneumonic plague and prophylaxis during epidemics.

Q-FEVER

Aetiology and epidemiology. Caused by *Coxiella burnetii*. Domestic sheep and cattle are the main animal reservoir. Humans are infected by: handling infected animals or meat; inhalation of contaminated dust; or drinking contaminated milk.

There are less than 100 cases per annum in the USA. At-risk groups include farmers, veterinary workers and laboratory staff involved in culture of *Coxiella*.

Clinical features. Incubation period usually 2–3 weeks. Symptoms and signs include fever, headache, anorexia, with primary atypical pneumonia, and, occasionally, infective endocarditis or hepatitis.

Laboratory diagnosis. By serology.

PSITTACOSIS

Aetiology and epidemiology. Caused by *Chlamydia psittaci*. The natural reservoir is birds who may have either a respiratory infection or be asymptomatic carriers. Humans are infected by inhaling air contaminated with the respiratory discharge of infected birds, or by direct contact.

Clinical features. Incubation period 1–4 weeks. Is an influenza-like illness with pyrexia, dry cough, occasionally with mucopurulent sputum and bilateral patchy consolidation of the lungs. Complications include myocarditis and endocarditis.

Laboratory diagnosis. Serology; culture is possible (culture is rarely performed because of the danger of laboratory-acquired infection).

Treatment and prevention. Tetracycline is the drug of choice. Contact with infected birds should be avoided.

TYPHUS AND RELATED SPOTTED FEVERS

Aetiology and epidemiology. Typhus is caused by various rickettsial species and is spread by ticks. Particular diseases are associated with different species and are found in different geographical locations (see p. 86). It is an infection associated with poor hygiene and overcrowding; epidemics may occur amongst refugees.

Clinical features. The incubation period is 7–14 days. An eschar may form at the site of the insect bite. There is high fever, a maculopapular rash and headache. The organism multiplies in the vascular endothelium and severe untreated infection results in haemorrhage and shock, with a high mortality. The severity of the clinical condition varies with the rickettsial species.

Laboratory diagnosis. By: serology; various immunoassays are available (e.g. complement fixation test; immunofluorescence tests).

Treatment and prevention. Tetracycline or chloramphenicol. Prevention involves reducing tick exposure and delousing of clothing/bedding. Killed vaccine for epidemic typhus is available for high-risk groups.

RABIES

Aetiology and epidemiology. This is an RNA virus present in the saliva of infected animals; man becomes infected following a bite from an infected animal. The organism spreads via the nerves to the CNS. It is found worldwide although the UK is currently rabies-free. Animal hosts include foxes, badgers, dogs, cats and bats.

Clinical features. The incubation period is 4–13 weeks, occasionally longer.

Sore throat, irritability, pyrexia, muscle spasms, convulsions and hydrophobia develop. Death results from cardiac or respiratory arrest.

Laboratory diagnosis. Immunofluorescence studies can be performed on skin biopsies; isolation of virus; RT-PCR of CSF.

Prevention
- Quarantine of imported animals (UK); animal vaccination in endemic areas.
- Post-exposure prophylaxis with rabies immunoglobulin following a bite from an animal with suspected rabies, and by active immunization.
- Vaccination of at-risk groups (veterinary workers; travellers to endemic areas).

LASSA FEVER AND OTHER VIRAL HAEMORRHAGIC FEVERS

Aetiology and epidemiology. Lassa fever is caused by an arenavirus and is endemic in West and Central Africa (300 000 cases per year, with 500 deaths). The animal reservoir is the rat. Humans are infected by contact with infected urine or saliva. Person-to-person spread can occur, which is associated with severe illness. Similar infections are caused by other arenaviruses (e.g. Marburg virus; Ebola virus) across Africa and Asia, with various animal reservoirs (e.g. monkeys).

Clinical features. Incubation period commonly 6–21 days. Fever, headache and myalgia, followed by severe multisystem infection develop. There is a high mortality (20–30%).

Laboratory diagnosis. By serology.

ORF

This is a disease of sheep, particularly lambs, caused by the orf virus. It infects sheep handlers, producing a maculo–papular lesion on the hand.

RAT BITE FEVER

This is a rare, febrile illness with abscesses and lymphadenopathy caused by *Spirillum minor* or *Streptobacillus moniliformis* (see p. 70). Humans become infected following a rat bite or through consumption of contaminated milk.

TOXOPLASMOSIS

Aetiology and epidemiology. Infection caused by *Toxoplasma gondii* (see Chapter 21, p. 126). Humans acquire the organism following contact with cats or soil contaminated with oocysts from cat faeces; the majority of infections are asymptomatic. Any organ can become infected, including the eye, liver, spleen, CNS and heart. Toxoplasmosis is a cause of glandular fever syndrome (pharyngitis, fever and malaise). Infection in pregnancy may result in congenital toxoplasmosis, resulting in abortion or stillbirth. It can cause encephalitis in the immunocompromised, particularly AIDS patients.

Laboratory diagnosis. Serology or histology of lymph nodes.

Treatment and prevention. Most cases do not require treatment. In severe infections, sulphonamide and pyrimethamine are used. The immunocompromised and pregnant women should avoid contact with cat faeces.

CHAPTER 45

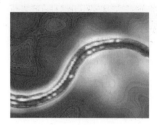

Pyrexia of Unknown Origin

Definition. A patient with a pyrexia lasting for 10 days or longer without an obvious infective cause.

Aetiology. Infections cause approximately 70% of acute pyrexias of unknown origin (PUO; <3 weeks) and approximately 30% of cases of chronic PUO (>3 weeks). Other causes include neoplasms, collagen diseases, endocrine abnormalities, drug reactions and malingering.

Infective causes of PUO

These can be classified according to site of infection or microbial aetiology.

SITE OF INFECTION

Infections which may present with a PUO include:
- endocarditis: due to many organisms, including viridans streptococci and *Staphylococcus epidermidis*;
- urinary tract infections: may present without urinary symptoms, such as dysuria and frequency, particularly in the elderly; occasionally involves the upper tract, with pyelonephritis;
- abscesses: dental, liver, spleen abdomen, pelvis, perinephric; causative organisms include coliforms, anaerobes, *Streptococcus milleri* and *S. aureus*;
- ear and upper respiratory tract infections in children; associated with *Haemophilus influenzae*, *S. pyogenes* and *S. pneumoniae*.

SPECIFIC CAUSES

Organisms which are commonly associated with PUO are shown in Table 45.1.

Clinical and laboratory diagnosis
- A careful history and full clinical examination are important in guiding appropriate laboratory investigations. Direct questioning may identify risk factors associated with certain infections (Table 45.2).

Table 45.1 Common microbial causes of PUO.

COMMON MICROBIAL CAUSES OF PUO	
Group	Organism
Bacteria	Mycobacterium tuberculosis Salmonella typhi Brucella
Viruses	Epstein–Barr virus Cytomegalovirus (CMV) Hepatitis A and B viruses
Parasites	Entameoba histolytica Plasmodium Pneumocystis carinii
Fungi	Cryptococcus Candida Aspergillus

Table 45.2 Risk factors and associated infections.

RISK FACTORS	
Risk factor	Associated infection
Travel	
Tropics	Tuberculosis, malaria, typhoid, amoebiasis
Mediterranean area (drinking unpasteurized milk)	Brucellosis
Pets	
Parrots	Psittacosis
Dogs	Toxocariasis
Cats	Toxoplasmosis
Occupation	
Sewer worker	Leptospirosis
Cattle/sheep farmer	Q-fever
Hobbies	
Water sports	Leptospirosis
Underlying conditions	
Diabetes mellitus	Abscesses
Immunocompromised	Consider atypical infections, e.g. M. avium-intracellulare in AIDS patients; cytomegalovirus infection in transplant patients
Recent surgery	Abscesses

PATTERNS OF FEVER

Pyrexia	Infection (examples)
Intermittent (returns to normal between peaks; peaks every 2–3 days)	Malaria
Undulating	Brucellosis
Swinging (frequent high peaks)	Abscesses
Low grade	Intravascular catheter-related

Table 45.3 Patterns of fever associated with infections.

INVESTIGATIONS FOR PUO

Investigation	Disease (examples)
Erythrocyte count	Anaemia associated with endocarditis
Leucocyte count: eosinophilia	Parasitic infection
Blood films: parasites	Malaria; trypanosomiasis
atypical mononuclear cells	Glandular fever
Blood cultures (prolonged culture)	Bacterial and fungal bacteraemia
Midstream urine: culture	Urinary tract infection
haematuria	Endocarditis
Early morning urine (three complete specimens)	Renal tuberculosis
Serology (as directed by history and examination)	Viral and bacterial (including chlamydial and rickettsial infections)
Liver function tests	Hepatitis
Imaging: X-rays, ultrasound scans, computerized tomography	Tuberculosis, osteomyelitis, abscesses, and other deep foci of infection
Biopsies: lymph nodes	Tuberculosis
bone marrow	Leishmaniasis
liver	Hepatitis; tuberculosis

Table 45.4 Investigations for PUO.

- Clinical examination should include an assessment of the pattern of fever (Table 45.3), examination of the ear, eye and oral cavity, a rectal examination and a search for rashes, bites and lymphadenopathy.
- Investigations: a range of non-specific and specific tests may be performed, both microbiological and non-microbiological (haematology, histology, imaging, clinical chemistry). Examples are shown in Table 45.4.
- Therapeutic trials may be considered for patients in whom a diagnosis has not been made; e.g. a patient recently arrived from the Indian subcontinent with a PUO might be considered for a trial of antituberculous therapy.

CHAPTER 46

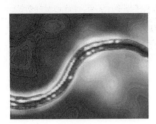

Epidemiology and Prevention of Infection

Community-acquired infections are acquired in the community not in hospitals. Hospital-acquired ('nosocomial') infections are those where the infecting organism is acquired during hospitalization. Normally, such infections are evident during the patient's hospital stay, but, occasionally (e.g. wound infection), the infection presents after the patient's discharge home. Studies in the UK and USA have estimated the incidence of hospital-acquired infection to be between 5 and 10% of all in-patients. The principal sites of infection are the urinary tract, wounds and lower respiratory tract; more recently, infections of intravascular catheters have become an important cause of hospital-acquired infection.

Increased risk of infection is associated with: prolonged hospitalization; intensive care; the use of invasive, prosthetic devices (e.g. intravenous catheters, urinary catheters, endotracheal tubes); an immunocompromised host, including those with impaired local host defences (e.g. burns, trauma). The widespread use of antibiotics in hospitals results in the selection of antibiotic-resistant organisms in the hospital environment; patients in hospital who receive antibiotics become colonized frequently with these antibiotic-resistant organisms, which may subsequently cause infection.

Strategies to reduce the risk of infection in individuals or in communities are dependent on an understanding of the sources and modes of transmission of microorganisms.

EPIDEMIOLOGY OF INFECTION

Epidemiology is the study of the aetiology and occurrence of infections in populations. Factors affecting the epidemiology of infections are shown in Fig. 46.1. Infections may be either sporadic, or occur in outbreaks, with two or more related patients, suggesting that transmission has occurred.

Definitions
- The *incidence rate* is the number of new cases of acute disease in a speci-

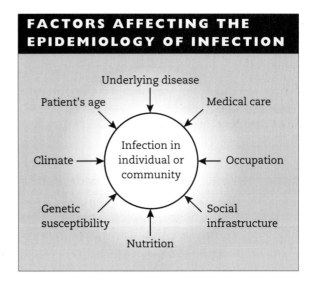

Fig. 46.1 Factors affecting the epidemiology of infection.

fied population over a defined period and is expressed as a proportion of the total population of the community involved.
- The *prevalence rate* is the total number of cases present in a defined population at a certain time point or over a certain period.
- The *attack rate* is the proportion of the total population at risk who became affected during an outbreak.
- Infections which remain present in the population are called *endemic*; an increase in incidence above the endemic level is described as hyperendemic.

SOURCES OF INFECTION

Endogenous. These are infections caused by organisms which are part of the host's normal flora.

Exogenous. These infections are caused by organisms acquired from other humans, animals or the environment.
- *Humans:* infections are commonly acquired from other humans who may be infected or carry an organism asymptomatically; carriers may be convalescent (i.e. recovered from the infection, but continuing to excrete the organism), or healthy carriers who have not suffered overt infection (e.g. health care workers).
- *Animals:* human infections acquired from animals are termed anthropozoonoses. Both domestic and wild animals may be sources of infection; the animal may be infected and carry the organism asymptomatically.

- *Environment:* large numbers of microorganisms are found in soil and water, but very few are associated with human infection (e.g. *Clostridium tetani*; *Aspergillus*, *Legionella*).

ROUTES OF SPREAD OF INFECTION

AIRBORNE

Airborne spread is by:
- respiratory droplets from an infected individual (e.g. influenza virus, *Neisseria meningitidis*, *Mycobacterium tuberculosis*);
- squamous epithelial cells (e.g. *S. aureus*);
- fine droplet spread from ventilator nebulizers (e.g. *Pseudomonas*); or from air-conditioning systems (e.g. *Legionella*);
- spores, e.g. *Aspergillus fumigatus* from contaminated air-handling plants.

CONTACT TRANSFER (Tables 46.1 and 46.2)

Sources of organisms (e.g. respiratory secretion, wound and other skin lesions, urine, faeces) are transferred to an individual either directly or via a vector such as the hands of staff or objects ('fomites'), e.g. surgical instruments, utensils, medical equipment. Other vectors include food, water and insects (e.g. mosquitoes, ticks). Many infections can also be spread via oral–oral or sexual contact. Faecal–oral transmission is a very important mechanism of spread of infection (Table 46.1). Trans-placental infection (via placenta from mother to fetus), e.g. rubella, can also occur.

ORAL TRANSMISSION	
Vector	**Infection (example)**
Direct	Shigellosis
	Rotavirus infection
Food	Salmonellosis
	Campylobacter enteritis
Milk	Brucellosis
	Tuberculosis
Water	Cholera
	Typhoid

Table 46.1 Examples of oral transmission.

TRANSMISSION BY DIRECT CONTACT

Contact	Infection
Skin	Dermatophyte infection Leprosy
Oral	Glandular fever
Sexual	Gonorrhoea Chlamydial infection HIV infection HBV infection
Inoculation	HBV infection HCV infection HIV infection

Table 46.2 Examples of transmission by direct contact or transfer.

FACTORS AFFECTING TRANSMISSION OF MICROORGANISMS

Survival in the environment

Microorganisms that spread between humans rely on various strategies.
- Respiratory viruses survive poorly outside the host; successful transmission requires production of large numbers of infectious virions. However, the number of virions sufficient to cause infection may be small.
- *Mycobacterium tuberculosis* can survive outside the host; transmission can occur with few bacteria.
- *Neisseria gonorrhoea* dies rapidly outside the host; therefore transmission is dependent on direct contact.

Establishment on new host

To attach and survive on a new host, an organism must combat the physical, chemical and immunological barriers which protect the host at each port of entry (e.g. respiratory tract, gastrointestinal tract, skin, urogenital tract). Organisms have a variety of virulence factors which promote attachment and survival (see Chapter 23).

STRATEGIES TO PREVENT INFECTIONS

- Reduce exposure to sources of infection:

- Prevention can only be achieved by knowledge and understanding of the source of the organism, the mechanism of transmission and the pathogenesis of infection. These principles can be applied when considering either community-acquired or hospital-acquired (nosocomial) infections and include:

    (a) reduce or eliminate sources of infection;

    (b) interrupt transmission;

    (c) prevent the establishment of infection following exposure to the pathogen by increasing host resistance (e.g. with prophylactic antibiotics, immunization).

Reducing sources of infection

Complete *elimination* of an infection worldwide is difficult and expensive, but has been achieved with smallpox and is aimed for poliomyelitis and measles viruses. Examples of strategies for reducing the source of infection are shown in Table 46.3. The decreased incidence of many infectious diseases in developed countries is associated with improved social conditions and reduced overcrowding; this has resulted in reduced transmission by aerosol, faecal–oral and direct contact transmission.

REDUCTION IN SOURCES OF INFECTION		
Source	Infection	Action
Human	Salmonellosis	Screening of food handlers for *Salmonella* carriage followed by isolation and treatment as appropriate
	Tuberculosis	Isolation; follow-up cases; contact tracing; investigation and treatment of contacts
Animal	Rabies	Quarantine; vaccination of pets
	Anthrax	Burning infected dead animals; vaccination of animals
Environment	Legionellosis	Improved design and maintenance of humidification systems and air conditioning systems, including chlorination

Table 46.3 Examples of reduction in sources of infection.

Preventing transmission of infection

AIRBORNE

Isolation of patients with infections which spread by respiratory droplets is feasible in only a limited number of conditions (e.g. tuberculosis). It is more important in preventing transmission within hospitals.

FAECAL-ORAL

Direct

- Good personal hygiene (handwashing).
- Isolate patients with enteric infections.

Water

- Separate sewage from drinking water.
- Provide potable water (filtration/chlorination) constantly monitored by microbiological testing.

Milk

- Ensure health of cattle.
- Ensure hygienic collection and storage of milk.
- Pasteurize or sterilize milk.
- Monitor microbiology of products.

Food

- Efficient refrigeration of certain foods.
- Separate storage and preparation areas for possible contaminated foods (e.g. uncooked poultry) from cooked produce.
- Thorough cooking and reheating procedures; complete defrosting before cooking.
- Well-designed catering facilities with high standards of personal hygiene; exclude from work any staff with infections (e.g. staphylococcal skin infections, diarrhoea, Salmonella carriers).
- Food contamination is an important public health problem, >25 000 notified cases of food poisoning occur per annum in the UK; many more cases are probably unreported.

DIRECT AND CONTACT TRANSFER

- Good personal hygiene.
- Sexually transmitted diseases can be prevented by: safe sex; contact tracing; and health education campaigns.

- Blood-borne virus infections (e.g. hepatitis viruses B, C, HIV) can be prevented by: sterilization of surgical instruments; safe sharps practice; provision of needles for intravenous drug abusers; and screening of blood and organ donors.

Increasing host resistance

IMPROVED GENERAL HEALTH

Malnutrition and parasitic infection are important factors that weaken host resistance to other infections.

ANTIMICROBIAL PROPHYLAXIS

- Surgical prophylaxis is administered mainly in hospitals. Prophylactic antibiotics are important in dental practice to prevent bacterial endocarditis.
- Long-term prophylactic antibiotics are prescribed in a limited number of conditions (e.g. post-splenectomy to prevent pneumococcal infection).
- Short-term prophylactic antibiotics (e.g. antimalarials) are given when travelling to endemic areas.
- Immunocompromised patients (e.g. transplant and AIDS patients) may receive long-term antimicrobial prophylaxis (e.g. acyclovir to prevent herpes simplex infections; co-trimoxazole to prevent pneumocystis pneumonia).

IMMUNIZATION

This is dealt with in Chapter 47.

NOSOCOMIAL INFECTIONS

Urinary tract infection

This is the most common nosocomial infection, and is often a complication of catheterization or urinary tract operative procedures. It is a particular problem in urology, obstetric, geriatric, spinal and intensive care units.

The source of the infecting organism may be endogenous or exogenous (cross-infection introduced during urological procedure, or manipulation of drainage system). Prevention includes the use of sterile instruments, closed drainage urinary catheter systems and careful aseptic technique during the introduction of urinary catheters and subsequent catheter care.

Wound infection

The incidence of wound infection varies according to the type of surgery (Table 46.4). Causative organisms depend on the site and type of operative procedure:
- *S. aureus*: the most common cause following surgery;
- coliforms and anaerobes: associated most commonly with 'contaminated' or 'infected' operations, mainly involving the gastrointestinal tract;
- coagulase-negative staphylococci: a frequent cause of device-associated infections;
- *Clostridium perfringens*: may result in gas gangrene, now rare in developed countries, but remains a problem in developing countries, particularly during wars; gas gangrene may be associated with lower limb amputation for vascular insufficiency;
- β-haemolytic streptococci: less common than staphylococcal infections, but may result in rapid local spread (cellulitis or fasciitis) and septicaemia.

Wound infections are frequently endogenous. Exogenous acquisition may be acquired intra-operatively or post-operatively, either by the airborne route or by direct contact (cross-infection).

Prevention
- Pre-operative measures:
    (a) decontamination of skin surfaces immediately pre-operatively;

WOUND INFECTION

Type of surgical wound	Example of surgery	Incidence of wound infection
'Clean': does not involve gastrointestinal, respiratory or genitourinary tracts	Orthopaedic or cardiac surgery	<5%
'Contaminated': involves site with normal flora (apart from skin)	Operations on gall bladder, genitourinary	20%
'Infected': operation site infected at time of surgery	Emergency surgery for burst, infected appendix	>50%

Table 46.4 Wound infection relative to surgery undertaken.

(b) prophylactic antibiotics for operations where there is a significant risk of wound infection.
- Intra-operative measures:
(a) modern operating theatres are designed to provide positive pressure ventilation with filtered air so that 'clean' air enters and passes to less clean areas; the ventilation systems reduce the number of microorganisms entering the theatres and, by directing the airflow away from the patient, the number of bacteria-carrying particles generated around any operation wound is minimized.
(b) the use of sterile instruments, careful handwashing techniques and the wearing of sterile surgical gloves;
(c) other procedures to prevent airborne transmission of organisms into open wounds include the wearing of headware and close-woven operation gowns;
(d) good surgical technique to reduce devitalization of tissue and haematoma formation.
- Post-operative measures:
(a) 'no-touch' techniques in the cleaning and dressing of wounds have been a major factor in reducing wound infection;
(b) the use of specially ventilated ward areas for dressing of extensive wounds, such as burns;
(c) isolation of patients with heavily infected wounds, particularly with antibiotic-resistant organisms.

Lower respiratory tract infection

This is an important cause of morbidity and mortality in hospitalized patients. It is often a complication of intubation (anaesthesia, intensive care) which bypasses the normal physical defences of the respiratory tract (mucous-trapping, ciliated epithelium).

Prevention. Many infections are endogenous and isolation of infected patients is of limited value. The use of single-use suction catheters and sterilized anaesthetic circuits with inspiratory and expiratory microbial filters reduces the chance of cross-infection.

Intravascular catheter infections

The use of both peripheral and central vascular catheters (CVC) is associated with sepsis, ranging from insertion site infections to septicaemia. Approximately 0.2% of peripheral devices and 4% of CVCs are associated with septicaemia. The source of sepsis is primarily from the patient's own skin flora, with organisms gaining access to the catheter either at the time of insertion or via connectors. The main organisms commonly causing catheter-related infection are: *S. epidermidis* (40–60%); *S. aureus* (20%); streptococci (10%); coliforms (10–15%); and *Candida* (<5%).

Treatment and prevention. This includes careful decontamination of skin around the catheter insertion site, aseptic techniques and daily monitoring of insertion sites. Catheters coated with antimicrobial polymers are being developed, which may help in preventing infection. CVC infections can be treated with antibiotics, such as vancomycin for *S. epidermidis*. Successful treatment rates vary from 50 to 80%.

Specific microorganisms associated with hospital-acquired infection

Certain organisms, e.g. multi-resistant strains of *Pseudomonas aeruginosa*, methicillin-resistant *S. aureus* (MRSA), and, more recently, vancomycin-resistant enterococci (VRE) need specific control strategies. MRSA is detailed below.

MRSA

- MRSA is resistant to methicillin which is equivalent to flucloxacillin.
- MRSA may also be resistant to other antibiotics, including gentamicin and fucidin.
- Epidemic strains of MRSA have been identified.
- MRSA can be carried in the following sites: anterior nares and pharynx; hair; hands; perineum and axilla; urine (particularly associated with catheters).
- If a patient is colonized with MRSA, eradication of the organism can be attempted, with: antiseptic ointment (e.g. mupirocin) to anterior nares and chlorhexidine baths.
- Infections with MRSA include: wound; urinary tract infection (catheter-related); pneumonia; and septicaemia.
- Treatment depends on sensitivity, but often includes vancomycin and trimethoprim.

ISOLATION PROCEDURES

Hospitals require a policy for isolating patients with certain transmissible infections (source isolation) and for isolating certain immunocompromised patients (e.g. bone marrow transplant patients) to prevent them

acquiring infection (protective isolation). Categories of isolation include:
- *strict isolation* for highly dangerous conditions, e.g. Lassa fever, viral haemorrhagic fever: these patients require treatment in designated isolation hospitals with special facilities (so-called category IV wards);
- *standard isolation* for conditions such as chickenpox, measles, infections with MRSA and other multiple antibiotic resistant organisms: includes single-room accommodation with closed door and *negative pressure* to outside; disposable gloves and aprons should be worn when attending the patient, and hands should be washed before leaving room; for enteric infections, procedures are required for safe disposal of faeces and urine.
- *Protective isolation*: facilities include single room accommodation, with *positive pressure* filtered air ventilation; gowns, aprons and masks need to be worn by all persons entering room; hands should be washed before entering.

INFECTION CONTROL TEAMS IN HOSPITALS

The team consists of infection control nurses and the infection control doctor. Their function is to:
- produce appropriate infection control policies (e.g. disinfection policy, isolation policy, needlestick injury policy);
- advise on infections in hospitals, including the need to isolate patients with specific infection or to close wards temporarily;
- liaise with the consultant in communicable disease control if infections have implications for the community.

CHAPTER 47

Immunization

The principle of immunization is to increase specific immunity to infection; this may be achieved by the administration of immune serum (passive immunization) or by administration of an antigen that primes the host immune system without causing disease (active immunization).

PASSIVE IMMUNIZATION

- Passive immunization (immunotherapy) gives immediate protection, but immunity lasts for a relatively short time (<6 months).
- Immunoglobulin for passive immunization is normally obtained from volunteers (e.g. individuals recently vaccinated) with high titres of specific antibody to a particular organism. However, for infections which are relatively common (e.g. hepatitis A) levels of specific antibody in the general population are high enough to allow the use of pooled plasma from normal blood donors (pooled human immunoglobulin).
- Indications for the use of passive immunization are limited and include:
    (a) post-exposure prophylaxis in immunocompetent hosts when immediate protection is required following exposure to the infection (tetanus, diphtheria, rabies, hepatitis B);
    (b) post-exposure prophylaxis in immunocompromised hosts (measles, varicella-zoster);
    (c) therapy (e.g. cytomegalovirus (CMV) infection in transplant patients and young children).

ACTIVE IMMUNIZATION

- The underlying principle is to stimulate the production of specific B- and T-lymphocytes (primary response) which develop into memory cells; subsequent exposure to the pathogen then results in the immediate and effective immune response (secondary response) (see Fig. 24.4).
- Active immunization provides long-lasting protection (often lifelong) against infection, but takes several weeks to become effective.
- Vaccination strategies are:

323

(a) to protect susceptible individuals against infection;

(b) to reduce the incidence of infection in the community (increase 'herd' immunity);

(c) to eliminate an infection in a particular country or worldwide (e.g. smallpox).

VACCINES

The ideal vaccine should be:
- able to induce an adequate and appropriate immune response without causing active infection;
- safe;
- inexpensive;
- stable;
- easy to administer.

Types of vaccine

LIVE ATTENUATED

- Live organism whose virulence has been attenuated; this is normally achieved by serial passage in artificial media, e.g. BCG (Bacille Calmette–Guérin) vaccine against tuberculosis; many viral vaccines.
- Advantages: replication of attenuated organism in body closely mimics true infection; generally, only one dose required; inexpensive.
- Complications: random mutations may occur very rarely, which result in revertence to virulence; cell lines used for virus production may become contaminated with other viruses (strict quality control procedures are necessary).
- Limitations: live vaccines cannot be given to immunocompromised patients or in pregnancy.

KILLED ORGANISMS

- Organism killed by chemical or heat treatment, e.g. vaccines against: rabies, influenza, polio (salk vaccine); *Salmonella typhi, Vibrio cholera, Bordetella pertussis, Yersinia pestis*.
- Advantage: can be given to immunocompromised; no revertence to virulence.
- Disadvantage: organism cannot multiply in tissues; several doses required for effective response; expensive.

TOXOIDS

- Inactivated (formaldehyde treatment) bacterial toxin vaccines e.g. tetanus, diphtheria.
- Disadvantage: relatively small molecules, therefore not immunogenic when given alone; require larger molecule (adjuvant) to produce effective immune response.

SUBCELLULAR OR SUBVIRAL FRACTIONS

- An alternative to using the whole organism is to identify and purify important subunit antigens for use as vaccines.
- Examples include the polysaccharide capsular vaccines for immunization against pneumococci, meningococci and *Haemophilus influenzae* type b; hepatitis B surface antigen, originally purified from the serum of chronic carriers, but now manufactured by recombinant DNA techniques; and influenza haemagglutinin (HA) and neuraminidase (NA) subunit vaccine.

INDIVIDUAL VACCINES

POLIOMYELITIS

- Both inactivated poliomyelitis vaccine and live attenuated vaccine (oral) are available.
- Incidence of polio drastically reduced in many countries since introduction of routine vaccination, eradication in the Americas attempted for the year 2000.

MEASLES/MUMPS/RUBELLA (MMR)

- Contains live attenuated measles, mumps and rubella viruses.
- Should be given to children irrespective of history of measles, mumps or rubella.

INFLUENZA

- Vaccine regularly updated to reflect current strains in circulation. Contains two type A and one type B virus strains; monovalent vaccine may be produced.
- Gives 70% protection; protects for approximately 1 year.
- Recommended for patients with chronic respiratory disease, heart disease, metabolic diseases, or immunosuppression. Recommendations for the elderly ($\geqslant 65$ years) differ by country.

HEPATITIS A
- Formaldehyde-inactivated vaccine made from a strain of hepatitis A virus.
- Recommended for travel to endemic areas.

HEPATITIS B
- Both active and passive immunization available:
    (a) vaccine: contains hepatitis B surface antigen prepared by recombinant DNA techniques; recommended for at-risk groups, e.g. health care workers and neonates born to HBV carrier mothers;
    (b) specific immunoglobulin for passive immunisation; recommended for post-exposure prophylaxis for non-immune individuals and neonates born to HBeAg carrier mothers.

HUMAN VARICELLA-ZOSTER IMMUNOGLOBULIN (VZIg)
- Prepared from pooled plasma of blood donors with recent chickenpox or herpes zoster.
- Recommended use for immunocompromised patients and infants <4 weeks of age without maternal antibody protection and non-immune pregnant women who have been in contact with cases of chickenpox, herpes zoster or newborns when mother develops chickenpox (1 week before or after delivery).

TUBERCULOSIS
- BCG contains live attenuated strain derived from *Mycobacterium bovis*.
- Should not be given to immunosuppressed patients.
- Routinely given in childhood in some countries.

STREPTOCOCCUS PNEUMONIAE
- A polyvalent vaccine containing purified capsular polysaccharide from 23 capsular types of pneumococci (accounts for 90% of isolates in the UK).
- Efficacy 60–70%; decreased efficacy in young children; not effective in children <2 years.
- Recommended for splenectomied patients.

CHOLERA
- Contains a mixture of two heat-killed serotypes.
- Protection lasts 3–6 months; recommended for travel to endemic areas.

TYPHOID

- Monovalent whole-cell typhoid vaccine; contains heat-killed *Salmonella typhi*.
- Recommended for travel to endemic areas.

MENINGOCOCCAL INFECTION

- A purified, heat-stable, lyophilized extract from the polysaccharide outer capsule of *Neisseria meningitidis*.
- Effective against serogroups A and C; no vaccine available yet for group B organisms.
- Recommended for close contact of cases of meningococcal infection.

DIPHTHERIA

- Toxin of *Corynebacterium diphtheriae*; rendered non-toxigenic (toxoid), but retaining antigenicity.
- Disease virtually eliminated in UK and other developed countries.

PERTUSSIS

- Suspension of killed *Bordetella pertussis*.
- Reduced number of cases per year; rarely associated with severe neurological illness in children.
- Newer vaccine preparations being developed.

CHILDHOOD IMMUNIZATION

Age	Vaccine
2–4 months (three doses at 4-week intervals)	Diphtheria, tetanus, pertussis (DTP) Oral polio *Haemophilus influenzae* type b (Hib)
12–18 months	Measles, mumps, rubella (MMR)
4–5 years (school entry)	Diphtheria, tetanus (DT) booster Oral polio booster
10–14 years	BCG (tuberculin-negative children only) Rubella (non-immune girls)
16–18 years (school leaving)	Oral polio Tetanus

Table 47.1 Childhood immunization schedule in the UK (1995).

TETANUS
- Prepared by treating a cell-free preparation of toxin with formaldehyde (tetanus toxoid).
- Part of childhood immunization schedules in many countries.
- Booster doses often given when patients present with contaminated wounds.

HAEMOPHILUS INFLUENZAE TYPE b (Hib)
- Capsular polysaccharide linked to protein.
- *H. influenzae* type b infection occurs principally in children <4 years; vaccination is normally given to infants during first 6 months of life.

ADMINISTRATION OF VACCINE

- Childhood immunization schedules are common in most developed countries and also in developing countries (WHO extended programme of immunization, EPI). Timings reflect the need to protect children against important infections early in childhood, the need to give booster doses for some vaccines (particularly toxoids), and the need to vaccinate at convenient times when take-up rates will be high. The UK immunization schedule (1995) is shown in Table 47.1.
- Adults require vaccination:
    (a) to boost childhood immunization (e.g. tetanus);
    (b) to cover travel to endemic areas (e.g. cholera, typhoid, yellow fever, polio, meningitis, hepatitis A);
    (c) for protection in high-risk occupations (e.g. hepatitis B in health care workers; rabies in laboratory workers);
    (d) for protection of at-risk groups (pneumococcal vaccine for postsplenectomy patients; influenza vaccine for elderly and patients with chronic respiratory/cardiac conditions).

CHAPTER 48

Sterilization and Disinfection

Definitions
- *Sterilization*: the process whereby all viable microorganisms, including spores, are removed or killed.
- *Disinfection*: the process whereby most, but not all, viable microorganisms are removed or killed.
- *Pasteurization*: the process used to eliminate pathogens in foods such as milk. Spores are unaffected.

STERILIZATION AND DISINFECTION

Methods of sterilization and disinfection employed in clinical practice are related to the material being treated. Sterilization and disinfection can be divided into either physical or chemical methods. The efficacy of these methods is influenced by the processing time, the presence of organic materials (e.g. blood) and the material being treated.

Dry heat

Recommended for sterilization of heat-tolerant articles; dry heat kills microorganisms by oxidation.
- *Flaming to red heat*: metal instruments, such as dental reflection mirrors, can be sterilized by direct heating. In an emergency, scalpels can be sterilized by dipping the blade in methylated spirit and burning off the spirit.
- *Hot air oven*: usually operates at 160°C for 1 h to sterilize items such as glassware; time must be allowed for all items in the oven to reach 160°C before the 1 h exposure is timed.
- *Incineration*: complete burning of material in an incinerator; this is used for the safe disposal of items such as contaminated dressings, pathology specimens and laboratory cultures.

Moist heat

Sterilization by moist heat is more rapid and efficient than dry heat; the presence of water causes protein denaturation resulting in disruption of cell membranes, and improves heat penetration.

- *Autoclaving* (Plate 64): autoclaves work on the same principle as the domestic pressure cooker: saturated steam is produced under pressure, resulting in temperatures >100°C; this ensures total destruction of all microorganisms, including spores. Contaminated items are exposed initially to steam under pressure and when the required temperature is reached, steam is evacuated to produce a vacuum and dry the load. A typical cycle would be 15 pounds/square inch at 121°C for 15 min. More recently it has been shown that for sterilization of equipment contaminated with the organism causing Creutzfeld–Jakob, a cycle of 134°C for 18 min is required.

Autoclaving is the most common method of sterilization in hospitals. To ensure efficient sterilization, the process must be carefully controlled:

(a) each sterilization cycle must be monitored by continuous recording of the temperature and steam pressure;

(b) items must be loaded into the autoclave to allow even and complete exposure to the steam; they should be covered to prevent recontamination after removal from the autoclave;

(c) efficacy of the sterilization can be tested by chemical or biological methods: Browne's sterilizer control tubes contain a chemical which changes colour when exposed to various temperatures; spore strips (*Bacillus stearothermophilus* spores are normally used) can be placed amongst the items being autoclaved; survival of the spores after autoclaving indicates a problem with the process.

- *Boiling water*: boiling water baths are still used in the clinical situation (e.g. cleaning of instruments). However, they cannot be relied upon to sterilize instruments, as spores will survive boiling.
- *Pasteurization*: this is used to destroy vegetative bacteria (e.g. *Mycobacteria*, *Salmonella*, *Brucella*) in milk by heating to 63°C for 30 min (Holder method) or 72°C for 20 s (Flash method). Spores survive pasteurization.

Filtration

Filtration can be used to remove microorganisms from fluids that cannot tolerate heat. However, filtration is slow and many viral particles can pass through filters. Filtration of air is important in operating theatres and in pharmacy departments where sterile solutions are being prepared. These filters remove the majority of microorganisms, but do not result in sterile

PROPERTIES OF DISINFECTANTS

Group	Example	Bactericidal Gram-negative	Bactericidal Gram-positive	Sporicidal	Fungicidal	Viricidal	Mycobactericidal	Inactivated by organic matter	Human tissue toxicity	Uses	Notes
Alcohols	70% ethyl alcohol	+	+	−	−	−	−	±	±	Skin antiseptic	100% alcohol ineffective; need addition of water to 70% v/v; rapid action
Aldehydes	Formaldehyde	+	+	+	+	+	+	±	+	Fumigation	Gas or aqueous forms
	Glutaraldehyde	+	+	+	+	+	+	±	+	Disinfection of fibreoptic endoscopes	Staff exposure to fumes should be minimized
Biguanides	Chlorhexidine	−	+	−	−	−	−	+	−	Hand wash; skin antiseptic	
Halogens	Hypochlorites Chlorine	+	+	±	+	+	−	+	±	General environmental cleaning; blood spills; treating water	Decreased activity on storage
	Iodine	+	+	±	+	+	−	+	±	With alcohol, used for skin preparation; hand wash and skin ulcers	
Phenolics	Phenol (carbolic acid)	±	+	−	−	−	+	+	+	Absorbed by rubber; too irritant for general use	
	Clear phenolics	±	+	−	−	−	+	−	+	Decontamination of floors, etc.	
	Hexachlorophane	−	+	−	−	−	−	−	−	Powder form for skin application, skin disinfection	
	Chloroxylenols (Dettol)	−	±	−	−	−	−	+	±		
Quaternary ammonium compounds	Cetrimide	±	+	−	±	−	−	+	−	Skin disinfection	Bacteriostatic
	Benzalkonium chloride	±	+	−	+	−	−	+	−	Preservative of topical preparation/antimicrobial plastic catheters	

+, yes; −, no; ±, intermediate.

Table 48.1 Properties of commonly used disinfectants.

air. Current guidance recommends a certain quality of air in some of these areas.

Radiation

Many bacteria are killed rapidly on exposure to radiation. Gamma radiation, usually from a cobalt-60 source, is being used extensively in the commercial sector for the sterilization of materials such as plastic disposable syringes. This process does not result in a rise in temperature of the materials, but the time required for sterilization can be as long as 48 h. The equipment and safety procedures required for sterilization by gamma radiation limit its use within hospitals.

Chemicals

A variety of chemicals are used to kill bacteria and their spores; these are classified into disinfectants or antiseptics:
- *disinfectant*: a chemical used to remove or kill most viable organisms present (Table 48.1);
- *antiseptic*: disinfectant used for skin cleaning; does not sterilize skin; less irritant compared to disinfectant.

Some disinfectants, e.g. glutaraldehyde, are used to sterilize equipment, such as fibreoptic endoscopes.

Gases, such as ethylene oxide, are used to sterilize single-use medical items; however, it is toxic and explosive. Formaldehyde gas is used to disinfect microbiology containment laboratories following contamination episodes.
- The choice of disinfectant for different hospital applications is made on the basis of their antimicrobial activity, inactivation by organic material, and toxicity. Most hospitals formulate a disinfection policy which specifies which agent, the appropriate concentration, and the correct contact time for different situations.

Subject Index

Page numbers in *italics* refer to figures, those in **bold** refer to tables.

α-toxin 38
abortion, spontaneous 103
abscess 36, 37, 202, 218–19, 245
   Actinomyces 46
   amoebic 245
   Bacteroides 71
   Eikenella corrodens 70
   flucloxacillin 180
   Nocardia 47
   pyrexia of unknown origin 308, **309**, **310**
   Staphylococcus aureus 155
acidogenic theory of dental caries 216
acinetobacters 69
acne 203
acquired immunodeficiency syndrome *see* AIDS
Actinomycetaceae 46–7, **229**
actinomycosis, oral 219
acyclovir **198**, 199–200
adult T-cell leukaemia 107
aerobic organisms 10, 34
   see also Gram-negative bacilli; Gram-negative bacteria
agglutination reactions 172, 176
AIDS 106, 231, 242, 288–9, 307
   see also HIV infection
alveolitis, allergic 118
amantidine 197–8
amikacin 193
aminoglycoside-modifying enzymes 189
aminoglycosides 182–3, 193–4
amoebae 122, **123**, 124
amoebiasis 124, 188, **309**
amoebic infection 211, 245
amphoterocin B 195–6
ampicillin 180
anaerobes 10, 11
   see also Gram-negative bacteria
animal bites 62, 207
anthrax 8, 44, **302**, 303, **316**
antibiotic resistance 16, 18, 189–91
   acinetobacters 69
   aminoglycosides 183
   E. coli 53
   β-lactam antibiotics 179
   mycobacteria 79, 80
   Pseudomonas aeruginosa 68
   S. epidermidis 28
   sulphonamides 185
   tuberculosis 233
antibiotics 177, **178**, 191–3
   acting on cell membrane 182
   destruction/inactivation 189

β-lactam 177–81
   prophylaxis 192, 318
   protein synthesis inhibitors 182–4
   semi-synthetic 177
   sensitivity testing 171
   synergistic 191
antibody 156, 163, **164**
   detection 173, *175*
   fluorescence-labelled 172–3, *174*
   labelling 173, 176
   response kinetics **166**
antifungal agents 194–6
antigen 51, 163
   detection 173, **174**
   soluble 156
antigen–antibody interactions 171–2, *174*–5, 176
antimicrobial agents *see* antibiotics
antimycobacterial agents 188
antiseptics 332
antiviral agents 197
aplastic crisis 103
arthritis 28, 48, 214–15
ascariasis 137, *138*, 139
aspergillosis 117–18, 194, 196
attachment 155, 160
attack rate of infection 313
autoclaving 330
autoinfection 142
autotrophs 9
azithromycin 183
azoles 194

B-lymphocytes 163, 164, 166
β-toxin 27
Bacille Calmette–Guérin (BCG) vaccine 79, 231, 324, 326
bacilli 20, **21**, **22**, 23
   aerobic 286, 297, 298
bacteraemia 270
bacterial endocarditis 28, 35, 36, 69, 154, 276
   see also endocarditis; infective endocarditis
bacterial slime 6
bacterial spores 7
bacterial toxin synthesis genes 18
bacteriophages 16, *17*, 18, 93–4, 190
   see also phage typing
bacteriuria, covert 250
balanitis 258
benzylpenicillin 179
biliary obstruction 298
blackwater fever 130
blastomycosis 194
boils *see* furuncles
bone infections 212–15
botulinus toxin 157

333

botulism 8, 40–1
brain abscess 268
bronchiectasis 67
bronchiolitis 230
bronchitis 230
brucella, intracellular survival 156
brucellosis 59–60, 301–2, **314**
  pyrexia of unknown origin **309**, **310**
bubonic plague 303–4
Burkitt's lymphoma 102, 295
burns 67, 68, 207, 297, **298**

campylobacter infection 63–4, 236–7
candidaemia 117
candidiasis 116, 192, 210, 218, 258
  antifungal agents 194, 195, 196
capsid, virus 88
carbuncles 202–3
cardiolipin antigen agglutination 74
catheters
  dialysis 297, **298**
  intravascular 320–1
  urinary 298, 318
ceftazidime 181
cefuroxime 181
cell division 9
cell envelope 1–3, 4, 5–7
cell wall 1–3, 4, 5
  synthesis inhibitors 177–81
cell wall-deficient bacteria 23–4
cellular immunodeficiency, infections 298–300
cellulitis 202, 204–5, 210
  Erysipelothrix rhusiopathiae 46
  Haemophilus influenzae infection 58
  Pasteurella infection 62, **302**
central nervous system infections 260, 266–9
cephalosporins 177, 178, 180–1
cervical cancer 111, 112
cervicitis 48, 255, 256
cestodes 144–8
Chagas' disease 132, 133–4
chancre 218, 257
chancroid 59, 258
chemotaxis 160
chest infection 68, 69, 181, **229**
  see also lung infection
chickenpox 101, 284, 292
Chinese liver fluke 150
chlamydiae 9, 82–5
chlamydial infection **173**
  conjunctivitis 210
  neonatal eye 287, 288
  non-specific urethritis 256
chloramphenicol 183–4, **192**
cholangitis 244–5
cholecystitis 244–5
cholera 65–6, 237–8
choriodoretinitis 210, 211
ciliates **123**, 125
ciprofloxacin 185
classification of bacteria 20, **21**, 22
clavulanic acid 180
clearance mechanism deficiencies 298
clindamycin 183
clostridia 38–40
clostridial infection **39**, 206, 207
clostridial toxin 38, 54
clotting cascade 157
co-agglutination reactions 172, 176

co-amoxiclav 192
coagulase 27
coccidia **123**, 125–31
cold sore 101, 219, 259
colitis 192, 238, 238–9
  see also enterocolitis; pseudomembranous colitis
collagenase 217
colony stimulating factor 167
commensals 154, 155, 220
common cold 96, 220–1
complement
  activation 52, 161, 163
  activation/deposition inhibition 156
  control 162
  deficiency 298
  fixation test 176
condylomata lata 257
congenital infections 281–5
congenital rubella syndrome 100
conjugation 17, 18
conjunctivitis 84, 107, 210
corneal ulcer 210
corynebacteria 42–4
cotrimoxazole 187
Creutzfeld–Jakob disease 268
croup *see* laryngotracheobronchiolitis
cryptococcosis 194, 195, 196
culture techniques 170–1
cystic fibrosis 67, 68, 298
cysticercosis 145–6
cytokines 160, 166, 167
cytomegalovirus infection 283, 294–5, **309**, 323
cytotoxins 125

defence mechanisms, specific 163–4, *165*, 166–7
delta agent 111
dental abscess 218–19
dental caries 35, 216
dental infections 216–18
deoxyribonuclease 32
dermatophytes 118–19, 194, 195
diagnostic methods 168–72
diaminopyrimidines 186–7
diarrhoeal disease 56
dihydrofolate reductase 189
diphtheria 43, 224–5
  toxin 18, 43, 157, 172
  toxoid 225
diphtheroid bacilli **21**, 43–4, **217**, **273**
disinfection 329–30, **331**, 332
disseminated intravascular coagulopathy 157, 270
DNA 7
  polymerase 14
duodenal ulceration 65
dysentery 55
  bacillary 234–5

ear infection 68, 118, 308
elastase 68
Elek test 43, 171, 172
elephantiasis 143
encephalitis 95, 266–7
  herpes simplex infection 101, 199, 293
encephalopathy with scrapie-like agents 268
endocarditis 28, 86, 180, 276–9, 308
  see also bacterial endocarditis; infective endocarditis
endotoxic shock 52, 157, 270
endotoxin 3, 156, 157
  campylobacter 63

Subject Index

E. coli 52
 Neisseria meningitidis 261
enoxacin 185
enteric fever 53, 54, 55, 235
enteritis
 Campylobacter 183, **302**, **312**
 protozoa 242–3
 Salmonella 53, 302
 viruses 241–2
Enterobacteriaceae 51–6, 299
enterococci 36
enterocolitis 54–5
enteroinvasive E. coli (EIEC) 238
enteropathic E. coli (EPEC) 238
enterotoxin 27
 cholera 237
 S. aureus 27
 staphylococcal 157
 vibrios 66
enterotoxogenic E. coli (ETEC) 238
enteroviral infection 95, 265, 280
envelope, virus 88
enzyme-linked immunosorbent assay (ELISA) 173, *175*
epidermo-dysplasia verruciformis 112
epididymitis 48, 84
epiglottitis 58, 223–4
erysipelas 202, 203–4
erysipeloid 46
erythrogenic toxin 18, 32
erythromycin 28, 182, 183
ethambutol 188
exanthema subitum 103, 296
 see also roseola infantum
exfoliative toxin 27
exotoxin 26–7, 52, 156, 157
eye infection 210–11
 Acanthamoeba spp. 124
 Chlamydia trachomatis 83, 84
 herpes simplex 293
 nematode 144
 neonatal 287–8
 Pseudomonas aeruginosa infection 68

famciclovir 200
fasciitis 202
Fc receptors 156, 163
fibrinolytic cascade 157
fifth disease 103
filamentous fungi 114
filtration 330
fimbriae 6–7, 48, 52
fish tapeworm 147
flagella 6, 64
flagellates **123**, 131–6
flora, normal 154, 202
 antibiotic depression 192
flucloxacillin 28, 179–80
fluconazole 194
flucytosine 195
flukes 148–53, 246
fluoroquinolones 185
folliculitis 202–3
food poisoning 234
 Bacillus cereus 45
 Clostridium botulinum 240–1
 Clostridium perfringens 239–40
 Staphylococcus aureus 240
 Vibrio parahaemolyticus 66, 240
 Yersinia enterocolitica 240

foscarnet 198–9
Fournier's gangrene 205
frequency dysuria syndrome 250
fucidin 188, **192**
fungal infection 114–21
 dermatophyte 209
 encephalitis **267**
 meningitis 265
 pericarditis 280, 281
 pyrexia of unknown origin **309**
 urinary tract **251**
fungi, characteristics 114
furuncles 202–3

ganciclovir **198**, 200
gangrene
 gas 38, 39, 206, 319
gas-liquid chromatography 176
gastric carcinoma 65
gastric ulceration 65
gastroenteritis 104, 107, 234
gastrointestinal infections 234
 bacterial 234–41
 protozoan 242–3
 viral 241–2
generation time 9
genetic engineering 18–19
genital herpes 101, 199, 259, 293
genital infection 83, 84, 254–9
genital ulcers 59
genital warts 259
genome, virus 88
genotypic variation 16, *17*, 18
gentamicin 193
Ghon focus 232
giardiasis 131, 188, 242
gingivitis 71, 217, 218
glandular fever syndrome 102, 221, 294–5
glomerulonephritis 32, 33, 276
glycocalyx 28
glycopeptides 181, 193
gonococcal infection, eye 287, 288
gonococcus 48
gonorrhoea 179, 254–5
Gram stain 169
Gram-negative bacilli **21**, 297
 aerobic 286, 297, 298
Gram-negative bacteria 2–3, 4, 20
 aerobic 181
 anaerobic 71–2
Gram-negative cocci 48–50
Gram-positive bacteria 1–2, 4, 20, **21**
granuloma
 inguinale 258
 tuberculous 79
griseofulvin 195
growth of bacteria 9–13
 growth curve 9, *10*
 growth factors 171
gummas 257

haemagglutinin 97, 98
haemolysins 27, 32, 156
α-haemolytic streptococci 30, 34–6
β-haemolytic streptococci 30–1, 31–4, 319
 Group A 18, 31–3, 155, 203, 204, 205
 burns 207
 septicaemia **272**
 surgical wound infection 208

venous ulcers 207
Group B 33, 203, **261**, **272**, 286
Group C 33, 207
Group G 34, 207
infective endocarditis 276–7
Lancefield grouping 171
neonatal meningitis 263, 264
orbital cellulitis 210
otitis media 225
pericarditis 280
pharyngitis 222
upper respiratory tract 220
wound infections 207
haemolytic uraemic syndrome 238
Haemophilus influenzae type b infection 57, 58, 59
haemorrhagic fevers, viral 306
Haverhill fever 70
Heaf test 79
helicobacters 64–5
hepatitis
  delta 110
  non-A, non-B 109, 249
  transfusion-related 249
  viral 108–10, 246–50
hepatitis A vaccine 109, 247, 326
hepatitis B infection 248–9
hepatitis B surface antigen (HBsAg) 326
hepatitis B vaccine 109, 248, 326
hepatocellular carcinoma 108
hepatomegaly, liver fluke 149, 150
herpangina 219
herpes 198, 199, 200
  genital 101, 199, 259, 293
  labialis *see* cold sore
  neonatal infection 288
  zoster 200
herpes simplex infection 170, 265
herpesvirus infection 92, 100–3
herpetic stomatitis 219
herpetic whitlow 293
heterotrophs 9
HIB vaccine 327, 328
histoplasmosis 120–1, 194, 280
HIV infection 105–7
  see also AIDS
hospitalization 312
host defence 156, 157
host resistance, increasing 317, 318
host response modification 197
host-parasite relationships 154–7
human bites 70
human body louse 76
human immunodeficiency virus infection *see* HIV infection
human retroviruses 105–7
human T-cell leukaemia virus 107, **267**
humoral immunodeficiency infections 298–300
hyaluronidase 27, 32
hydatid cyst **145**, 148, 246
hydrops fetalis 103

IgG antibody detection 176
IgM antibody detection 176
ileitis, terminal 56
imidazole antifungals 182
immune defences
  non-specific 158–63
  specific 163–7
immune status testing 176
immunization 323

active 323–4
childhood 327, 328
passive 323
immunocompromised patients 297–300
  Actinomyces infection 46
  Aspergillus infection 117–18
  coccidia 125, 126, 127, 128
  Coccidioides immitis infection 120
  Cryptosporidium parvum 242
  cytomegalovirus 102, 294
  disseminated varicella-zoster infection 101
  fluconazole 194
  fungal meningitis 265
  ganciclovir 200
  herpes simplex 101, 293
  HIV infection 106
  human papilloma virus 112
  Listeria monocytogenes infection 45
  Nocardia infection 47
  pneumonia 229–30
  Pseudomonas aeruginosa 67
  toxoplasmosis 307
  varicella-zoster 293
immunoglobulin 163, *165*, 323
immunological detection methods 170
immunomodulators 197
impetigo 202, 203
incidence rate of infection 312–13
inclusion body 82, 102, 112
inclusion granules 7
infection 155
  airborne 314, 317
  control team 322
  isolation procedures 321–2
  strategies for prevention 316–18
  transmission 315
infective endocarditis 276–9
  see also bacterial endocarditis; endocarditis
influenza 221–2, 314
  vaccine 221, 324, 325
influenza virus
  antigen detection **173**
  haemagglutination 89
interferons 167, **198**, 199
interleukins 167
intra-abdominal sepsis 71
intracellular killing 160
intracellular structures 7
intracellular survival 156
intravascular coagulopathy 52
  see also disseminated intravascular coagulopathy
isolation procedures 321–2
isoniazid 188, **192**
itraconazole 194

joint infections 212–15

kala-azar 135
Kaposi's sarcoma 103
keratoconjunctivitis 101
ketoconazole 195
kidney stones 298
klebsiellae 53
Koplik spots 216, 292
Kuru 170

L-forms 23, 87
laboratory media 12–13
β-lactam antibiotics 177–81

combination with aminoglycosides 183
  resistance 189
β-lactamase 50, 179
  Bacteroides production 72
  Haemophilus influenzae production 59
  klebsiellae production 53
lactobacilli, dental caries 216
laryngotracheobronchiolitis 222
Lassa fever **302**, 306
latent infection, viral 92
leg ulcers 67, 68
legionellae 156, 183
legionellosis 60–1, **316**
leishmaniasis 135–6
Lentivirinae 105
leprosy 80
leptospirosis 76–7, 302–3, **309**
leucocidins 27, 32, 156
lincomycin 183
lincosamides 183
lipase 27
lipopolysaccharides 2, 3, 4, 270
  complement activation 161
  endotoxic shock 157
lipoteichoic acid 2, 4, 32, 155
liquid growth media 11–12
listeria, intracellular survival 156
listeriosis 45, 285
liver abscess 245
liver fluke 148–50, 246
lower respiratory tract infection 226–33, 320
lung fluke 150–1
lung infection 28
  Bacteroides infection 71
  Cryptococcus neoformans 116
  influenza 221–2
  mucormycosis 119
  Pneumocystis carinii 127
  viral 221–2, 227, 228, 230
Lyell's disease 208
Lyme disease 75, 76
lymphocytes 163–4
lymphogranuloma venereum 83, 84, 259

M-proteins 32, 33
McConkey agar 12
macrolides 183
macrophages 159–60, 166
malaria **128**, 130, 131, 285
  pyrexia of unknown origin **309**, **310**
Mantoux test 79
measles 98, 99, 291–2
  antigen detection **173**
  encephalitis **267**
  Koplik spots 216
  passive immunization 323
  vaccination 99
  virus 316
memory cells 164, 166
meningitis
  antibiotic therapy **264**
  bacterial 260–3
  chloramphenicol 184
  Coccidioides immitis infection 120
  E. coli K1 antigen 52
  enteroviral 95
  fungal 265
  Haemophilus influenzae type b 59, 261–2
  herpes simplex 293

leptospirosis 77
Listeria monocytogenes 45, 180, 234
lymphocytic 23
meningococcal 50, 179, 263
  Neisseria meningitidis infection 155, 260–1
neonatal 263–4
pneumococcal 34, 262
post-operative 264
protozoan 266
rifampicin prophylaxis 187
streptococcal 34, 262
tuberculous 265
viral 265
meningococcal infection 49–50
  vaccine 325, 327
meningococcus 20
meningoencephalitis **267**, 288
mesenteric adenitis 56
mesophiles 11
mesosomes 1
messenger RNA *see* mRNA
metazoa 137–53
methicillin-resistant *Staphylococcus aureus* 28, 189, 321
metronidazole 187–8
miconazole 195
microaerophilic bacteria 10
microbial identification 171–2
microbial strategies 155–7
microscopy
  direct 169
  electron 170
moist heat sterilization 329–30
mononucleosis, infectious 103
  see also glandular fever syndrome
mosquitoes
  malaria transmission 128, 129, 130, 131
  nematode transmission 143–4
moulds 114
mRNA 14, *15*
mucormycosis 119, 196
mumps 98, 99, 291
  encephalitis **267**
  meningitis 265
  viral arthritis 215
mumps/measles/rubella (MMR) vaccine 291, 292, 294, 325
mutation 16
mycelia 114
mycobacteria 78–81, 156, 176
  atypical 23, 78, 80, 81, **229**
mycoplasmas 86–7, 183
mycoses 114, **115, 116**
myocarditis 95, 279
myonecrosis, clostridial 206

Nagler reaction 38, 171
nalidixic acid 185
nasopharyngeal carcinoma 103, 295
Negri bodies 112
neisseriae 5, **22**
nematodes 137, **138**, 139–44, 211
neonatal infection 281, 285–9
neuraminidase 97, 98, 217
neurosyphilis 257
neurotoxin 39, 40
nitrofurantoin 188
nitrogen source for growth 9
nitroimidazole 187–8
norfloxacin 185

Norwalk agent 242
nosocomial infection 312, 321
    acinetobacters 69
    klebsiellae 53
    Pseudomonas aeruginosa 67, 68
NSU *see* urethritis, non-specific
nucleic acid replication 90
nucleic acid synthesis inhibition 184–8
nutrient broth/agar 12

obstetric infections 289–90
ofloxacin 184
onchocerciasis 211
ophthalmic zoster 211
opsonization 156, *161*, 162
oral infection 71, 72, 218
oral mucosal ulceration 69
orf 306
organ transplantation **299**, 300
osmotic pressure 5
osteomyelitis 28
    Coccidioides immitis infection 120
    Haemophilus influenzae infection 58
otitis
    externa 225
    media 34, 58, 225
oxygen requirements 10–11

parainfluenza virus infection 98, 99, **173**
paralysis, *Clostridium botulinum* 234
parasites
    ecology 154
    liver 246
    metazoan 122, **123**, 137–53
    protozoan 122
    pyrexia of unknown origin **309**
    urinary tract infection **251**
paratyphoid fever 53, 235
parvobacteria 57–62
parvovirus infection, congenital 284
pasteurellosis 62
pasteurization 330
pathogenic microorganisms 154–5
pathogens, opportunistic **300**
pelvic inflammatory disease 48, 84
penicillin 36, 177, 178, 179, **192**
penicillin binding proteins (PBPs) 178, 179
peptidoglycan 1, 4, 161
peptidyl transferase 14
peptostreptococci, saliva **217**
pericarditis 280
perihepatitis 84
periodontal disease 217
periodontitis, juvenile 218
peritonitis 181
pertussis 61–2, 231
    vaccine 327
pH of medium 11
phage typing 28, 93
phagocytes 156, 159–60, *161*
phagocytic activity defects 298
phagocytosis 160, *161*
phagosome 160, *161*
phagosome–lysosome fusion inhibition 79, 83
pharyngitis 221, 222–3
pharyngotracheitis 107
photochromogens 80
physical defences 158, **159**
pinta 73, 75

piperacillin 180
pityriasis versicolor 117, 210
plague 56, 303–4
plaque 216
plasma cells 164
plasmids 7, *17*, 18, 69, 190
plasmodia 128–31
pleurodynia 95
pneumococcal pneumonia 179
pneumococci 5, 20, 34
    meningitis 262
    neonatal eye infection 287, 288
    penicillin-resistant 189
    vaccination 325, 326
pneumonia 28, 155, 226
    Bacteroides infection 71
    Chlamydia pneumoniae 85
    flucloxacillin 180
    Haemophilus influenzae infection 58
    Pneumocystis carinii 187
    viral **227**, 228
pneumonitis, neonatal 84
point mutations 16
poliomyelitis 95
    vaccine 324, 325
polyarthritis, reactive 56
polyene antifungals 182
polymerase chain reaction 176
polymorphs 160
polymyxins 182
polysaccharides 5
    slime 6
Pontiac fever 61
precipitation reactions 172
pregnancy
    genital herpes 259
    hepatitis E virus 111
    Listeria monocytogenes infection 45
    rubella virus 100
    see also congenital infections
prevalence rate of infection 313
prosthetic valves/joints 297, **298**
protease 68, 156
protein A 27
protein synthesis 14, *15*
    inhibitors 182–4
proteinase 27, 217
proteolytic theory of dental caries 216
protoplasts 5, 24, 87
Protozoa, classification 122, **123**
protozoan infection 124–36, 266, **267**
pseudohyphae 114
pseudomembranous colitis 40, 181, 183
pseudomonads 67–8
psittacosis 84–5, 277, 304–5, **309**
psychrophiles 11
puerperal sepsis 289–90
pyelonephritis, acute bacterial 250
pyrazinamide 188
pyrexia of unknown origin 308, **309–10**, 311

Q-fever 85, 86, 277, **302**, 304, **309**
quinolones 184–5

rabies 112–13, **267**, **302**, 305–6
    passive immunization 323
    reduction in source of infection **316**
    vaccine 324
radiation 332

Subject Index 339

rat bite fever 70, **302**, 306
rat flea 303
recombination 16, *17*, 18
Reiter's syndrome 84
relapsing fever 23, 75–6
replication 156
  inhibition 197
  viral 89, *91*, 197
resistance
  viral 197
  see also antibiotic resistance
respiratory syncytial virus 98, 99, **173**, 201
respiratory tract infection 85
  Chlamydia trachomatis 83
  enteroviral 95
  influenza virus 97
  Mycoplasma infection 87
  see also lower respiratory tract infection; upper respiratory tract infection
retroviruses 105–7, 200, **267**
rheumatic fever 32
rhinocerebral infection 119, 221
ribavirin **198**, 201
ribosomes 7, 14
rickettsiaceae 9, 85–6, 183
rifampicin 187, 188, **192**
rifamycins 187
rimantidine 197–8
ringworm fungi 118
Ritter's disease 208
RNA 14
roseola infantum 103
  see also exanthema subitum
rotavirus infection transmission **314**
rubella virus infection 99–100, 294
  congenital 100, 282–3, 294
  vaccination 100, 282–3

salivary duct obstruction 298
Salk vaccine 324
salmonellae 12, 13, **235**
  speciation 171, 172
salmonellosis 235–6, **314**, **316**
scalded skin syndrome, staphylococcal 27, 208–9, 287
scarlet fever 202, 208
schistosomes 148, 151–3, 246
schistosomiasis 151–3
scotochromogens 80
scrapie-like agents 268
septic arthritis 58
septic shock 157, 270
septicaemia 48, 180, 181, 270, **272**, **273**, **274**
  anthrax complication 44
  Listeria monocytogenes 45
  meningococcal 50
  pneumococcal 34
  Pseudomonas aeruginosa infection 68
  streptococcal neonatal 33
serology 173, 176
sexually transmitted disease 254–9, 317
shellfish
  contaminated 66, 246
  lung fluke infection 150
shigella 54
shigellosis 55–6, 234–5, **314**
shingles 101, 292–3
shunt infection 264
sialoadenitis, acute 218
sinusitis 34, 225

sixth disease 103
skin
  candidiasis 210
  dermatophyte fungal infections 209
  Staphylococcus aureus infection 155
  toxin-mediated conditions 208–9
sleeping sickness 132
  East African 133
  West African 132–3
slime layer 6, 28
soft tissue, pathogenesis of infections 202–11
solid growth media 12
specimens
  collection 168
  culture 170
spheroplasts 24, 87
spiral bacteria 22–3
spiramycin 183
Spirochaetales 73–7
spleen function impairment 34
spotted fevers 85, **86**, 305
staphylococcal infection 5, 25–7, 28
  coagulase-negative 271, 287, 297, **300**, 319
  fucidin 188
  scalded skin syndrome 208–9
staphylokinase 27, 230
sterilization 329–30, **331**, 332
stomatitis, denture 218
stomatitis, herpetic 101, 219, 293
streptobacillary fever 70
streptococcal infection 179, 321
streptococci 5, 30–6
  anaerobic 37
  gangrene 205
  group A **173**
  group B **173**, 255
  see also viridans streptococci
streptokinase 32
streptolysins 32
sulphadiazine 185
sulphadimidine 185
sulphonamides 185–6, 190, **192**
superinfection, staphylococcal of cellulitis 205
symbiosis 154
syphilis 23, 73–5, 179, 256–7
  congenital 74, 257, 284–5
  primary 218, 257, 285
  secondary 257, 285
  tertiary 257
systemic lupus 163

T-lymphocytes 164, **165**, 166
tapeworms 144–8
taxonomy of bacteria 20
teichoic acid 2, 4
  see also lipoteichoic acid
tetanus 8, 40, 269
  passive immunization 323
  toxin 157
tetracyclines 184
thermophiles 11
thrush 116
tick-borne disease 76, **86**, 305
tinea 118, **209**
tobramycin 193
tonsillitis, viral 221
toxic epidermal necrolysis 208
toxic shock syndrome 27, 209
toxigenicity 156–7

toxin
  A and B of *Clostridium difficile* 239
  bacterial in skin changes 202
  bacteriophage coding 93
  production by *S. aureus* 26, 27
toxocariasis 140, **267**, **302**, **309**
toxoplasmosis **302**, 307, **309**
  congenital 211, 283–4, 307
trachoma 83, 84, 211
transcription 14
transduction 16, *17*
transfer RNA *see* tRNA
transformation 16, *17*
translation 14
transmission 155, 315
  faecal–oral 314, 316, 317
  prevention 317
transplacental spread 314
transposons 18, 190
trematodes 148–53
trichomoniasis 188
trimethoprim 186–7, 189
tRNA 14, *15*
tropical spastic paraparesis 107
trypanosomiasis 132, **267**
trypomastigotes 132, 133
tsetse fly 132, 133
tubercle bacillus 23, 76
tuberculosis 78, 79, 231–3
  congenital 285
  pyrexia of unknown origin **309**
  reduction in source of infection **316**
  rifampicin 187
  transmission **314**
  vaccine 324, 326
  see also osteomyelitis, tuberculous
tumour necrosis factor 167
typhoid fever 53, 235
  chloramphenicol 184
  pyrexia of unknown origin **309**
  transmission **314**
  vaccine 327
typhus 85, 305

ulcers 206–7
  corneal 211
  duodenal 65
  genital 59
  leg 67, 68
  oral mucosa 69
upper respiratory tract infection 308
urease 65
urethritis 48, 255, 256
  Mycoplasma infection 87
  non-specific 215, 255–6
  Ureaplasma infection 87
urinary tract, immune defence 250–1
urinary tract infection 250–3, 318
  Proteus spp. 56

Pseudomonas aeruginosa infection 68
pyrexia of unknown origin 308
uterus, *Actinomyces* infection 46

vaccination 43
vaccines 325–8
vaginal infection 132
vaginosis 70, 71, 258
vancomycin 28, 181, **192**, 193
vancomycin-resistant enterococci (VRE) 36, 321
varicella-zoster 170, 323, 326
venous ulcers 206–7
verotoxogenic *E. coli* (VTEC) 238
vibrios 65–6
vidarabine **198**, 199
Vincent's angina 23, 72, 224
viral infection 318
  pericarditis 280, 281
  pyrexia of unknown origin **309**
  upper respiratory tract 220–2
viridans streptococci 30, 34, 35–6, 154, **217**, **272**, **278**, 308
virion 88
virulence factors 5, 6, 7, 163
  Bordetella pertussis 61
  establishment on new host 315
  Group A β-haemolytic streptococci 32
  Group B β-haemolytic streptococci 33
  viridans streptococci 35
virus 88–92
  classification **92**, 94
  drug resistance 197
  inactivation 197
  influenza 97–8
  replication 89, 91, 197
  urinary tract infection **251**
vulvovaginitis 258

warts
  genital 259
  human papilloma virus (HPV) 112
Waterhouse–Friderichsen syndrome 261
Weil's disease 23, 77
whipworm 140
whooping cough *see* pertussis
winter vomiting disease 242
wound infections 36, 207, 319–20

yaws 23, 73, 75
yeast-like fungi 114
yeasts 114
  antifungal agents 195
  Cryptococcus neoformans 115–16
  non-specific urethritis 255

zidovudine **198**, 200
Ziehl–Neelsen's stain 169, 170
zoonoses 301–7
zygomycosis *see* mucormycosis

Organism Index

Page numbers in *italics* refer to figures, those in **bold** refer to tables.

Absidia corymbifera 119
Acanthamoeba spp. 122, **123**, 124, 211
Acinetobacter spp. 69
Actinomyces israelii (*A. israelii*) 21, 46, **213**, 219
Actinomyces spp. 46, 216
adenovirus 107–8, 200, 221, **227**, 242, **251**
Ancylostoma duodenale 141–2
Anopheles spp. 128, 130
arenavirus **302**, 306
Ascaris lumbricoides 137, **138**, 139
Aspergillus fumigatus (*A. fumigatus*) **116**, 117–18, 196, **299**, 313
Aspergillus niger (*A. niger*) 117–18
Aspergillus spp. 114, 194, **229**, 264, 280, **300**, **309**, 313
astrovirus 104–5, 242

Bacillus anthracis (*B. anthracis*) 2, 7, **21**, 44, **302**, 303
Bacillus cereus (*B. cereus*) **21**, 45, **234**
Bacillus spp. 8, 20
Bacillus stearothermophilus 330
Bacteroides fragilis (*B. fragilis*) **22**, 71, 245
Bacteroides spp. 71, 205, **213**, **217**, 290
Balantidium coli (*B. coli*) **123**, 125
Blastomyces dermatitidis (*B. dermatitidis*) **116**, 120
Blastomyces spp. 114, 196
Bordetella pertussis (*B. pertussis*) **22**, **58**, 61–2, 324, 327
Bordetella spp. 183, 231
Borrelia burgdorferi (*B. burgdorferi*) **74**, 75, 76, **267**
Borrelia duttoni (*B. duttoni*) 75–6
Borrelia recurrentis (*B. recurrentis*) 2, 23, **74**, 75–6
Borrelia vincenti 23
Branhamella catarrhalis see *Moraxella catarrhalis* (*M. catarrhalis*)
Brucella abortus (*B. abortus*) **22**, **58**, 59–60, 301
Brucella melitensis (*B. melitensis*) **58**, 59–60, 301
Brucella spp. **309**, 330
Brucella suis (*B. suis*) **58**, 59–60, 301

calicivirus 104–5, 242
Calymmatobacterium granulomatis 258
Campylobacter coli (*C. coli*) 63
Campylobacter fetus (*C. fetus*) 63
Campylobacter jejuni (*C. jejuni*) 10, 11, **22**, 63, **234**, 236
Campylobacter pyloridis see *Helicobacter pylori* (*H. pylori*)
Campylobacter spp. **302**
Candida albicans (*C. albicans*) 114, **115**, 116, **251**, **254**, 258, **278**
Candida parapsillosis (*C. parapsillosis*) 116
Candida spp. 194, 195, **229**, 264, 280, 297, **298**, **299**, **309**, 321

Candida tropicalis (*C. tropicalis*) 116
Capnocytophaga spp. 69
Cardiobacterium hominis (*C. hominis*) 69
Chlamydia pneumoniae (*C. pneumoniae*) 82, 85, 228
Chlamydia psittaci (*C. psittaci*) 82–3, 84–5, 228, 277, **279**, 304–5
Chlamydia spp. **173**
Chlamydia trachomatis (*C. trachomatis*) 2, 82, 83–4, 170, 211, 215, 255, 256, 259, 277, 287
Cholera vulnificus 204
Chromobacterium violaceum 69
Chrysops spp. 144
Clonorchis sinensis see *Opisthorchis sinensis*
Clostridium botulinum (*C. botulinum*) 8, **21**, 38, 40, 157, 234, **234**, 241
Clostridium difficile (*C. difficile*) **21**, 38, 40, 181, 192, 239
Clostridium perfringens (*C. perfringens*) 2, 7, **21**, 38–9, 157, 171, 206, 213, **234**, 239, **273**, 290, 319
Clostridium spp. 8
Clostridium tetani (*C. tetani*) 7, 10, **21**, 38, 39–40, 157, 268–9, 313
Coccidioides immitis (*C. immitis*) **116**, 119–20
Coccidioides spp. 114, 196
coronavirus 96, 221, 230
Corynebacterium diphtheriae (*C. diphtheriae*) 7, **21**, 42–3, 93, 157, 171, 224, 327
Corynebacterium jeikeium (*C. jeikeium*) 43
Corynebacterium ulcerans (*C. ulcerans*) 42, 43
Coxiella burnetii (*C. burnetii*) 86, 228, 277, **279**, **302**, 304
coxsackie A virus 95, **97**, 219
coxsackie B virus 95, **97**
coxsackie virus 221, 265, 280
Cryptococcus neoformans 114, 115–16, 194, 195, 196, 264, **267**
Cryptococcus spp. **309**
Cryptosporidium parvum **123**, 125, 242
cytomegalovirus 102, 198, 199, 200, 210, 221, **227**, **229**, **267**, 280, **282**, 283, 294–5, **300**, **309**

Diphyllobothrium latum **145**, 147

Ebola virus 306
Echinococcus granulosus **145**, 147–8, 246
echovirus 95, 221, 265, 280
Eikenella corrodens (*E. corrodens*) 70
Endolimax nana 122
Entamoeba coli 122
Entamoeba histolytica (*E. histolytica*) 122, **123**, 124, 243, **309**
Enterobacter cholaceae (*E. cholaceae*) **22**
Enterobacter spp. 51, **52**, 53, 250
Enterobius vermicularis 137, **138**, 139–40
Enterococcus faecalis (*E. faecalis*) 36, 180, **251**
Enterococcus faecium (*E. faecium*) 36
Enterococcus spp. 30

Organism Index

Epidermophyton floccosum (*E. floccosum*)   118
Epidermophyton spp.   **115**
Epstein–Barr virus   102–3, 199, 221, 280, 295–6, **300**, **309**
Erysipelothrix rhusiopathiae (*E. rhusiopathiae*)   46, 204, 205
Escherichia coli (*E. coli*)   2, 9, 11, 12, 20, **22**, 51–3, 181, 202, 229, 238–9, 244, 245, 250, 251, **261**, 263, 264, 271, 276, 286, 290
Escherichia spp.   51

Fasciola hepatica (*F. hepatica*)   148–9
Fasciola spp.   246
Fusobacterium fusiformis   218
Fusobacterium necrophorum (*F. necrophorum*)   72
Fusobacterium nuciliarsum   217
Fusobacterium nucleatum (*F. nucleatum*)   72
Fusobacterium spp.   **22**

Gardnerella vaginalis   70, 255, 258
Giardia lamblia   **123**, 131, 188, 242–3

Haemophilus ducreyi (*H. ducreyi*)   **58**, 59, 258
Haemophilus influenzae (*H. influenzae*)   **22**, 57–9, 171, **173**, 180, 183, 187, 188, 204, 205, 210, 227–8, 229, 230, 231, 287, 288, 298, 308
Haemophilus influenzae type B   **213**, 214, 223, 224, 225, 261–2, 262, **264**, 325, 328
Haemophilus parainfluenzae (*H. parainfluenzae*)   **22**, **58**, 59
Haemophilus spp.   183, 220
Helicobacter pylori (*H. pylori*)   **22**, 64–5
Hepadnaviridae   108
hepatitis A virus (HAV; enterovirus 72)   109–10, 246–7, **309**
hepatitis B virus (HBV)   91–2, 108–9, 176, 199, 247–9, 289, **300**, **309**
hepatitis C virus (HCV)   110, 176, 199, 249
hepatitis D virus (HDV)   110, 249
hepatitis E virus (HEV)   111, 249
hepatitis G virus (HGV)   110
herpes simplex virus (HSV)   100–1, 198, 199, 210, 219, 221, **227**, **229**, **254**, 259, 266–7, **267**, 288, 293–4
Histoplasma capsulatum (*H. capsulatum*)   114, **116**, 120, 194, 196
HTLV-1   107, **267**
human herpes virus 6 (HHV 6)   103, 296
human herpes virus 7 (HHV 7)   103
human herpes virus 8 (HHV 8)   103
human immunodeficiency virus (HIV)   92, 105–6, 176, 199, 200, 233, **254**, 255, **267**, **282**, 288–9
human papillomavirus (HPV)   111–12, 259
human polyoma virus   **251**
human T-cell leukaemia virus see HTLV-1
Hymenolepis nana   **145**, 148

influenza A virus   198, 280
influenza virus   97–8, 221, **227**, 230, 313
Isospora belli   **123**, 126

Klebsiella aerogenes (*K. aerogenes*)   53
Klebsiella oxytoca (*K. oxytoca*)   53
Klebsiella pneumoniae (*K. pneumoniae*)   **22**, 53
Klebsiella spp.   51, **52**, 229, 244, 250, 286

Lactobacillus acidophilus   10
Lactobacillus spp.   **21**
Legionella pneumophila (*L. pneumophila*)   **22**, **58**, 60–1, 184, 226, 228, **229**, 230
Legionella spp.   **173**, 230, 313

Leishmania braziliensis (*L. braziliensis*)   134, 136, **137**
Leishmania donovani   134, 135
Leishmania major (*L. major*)   134, 135–6
Leishmania spp.   **123**, 131, 134–6
Leishmania tropica (*L. tropica*)   134, 135–6
Leptospira biflexa (*L. biflexa*)   76
Leptospira canicola (*L. canicola*)   23
Leptospira icterohaemorrhagiae (*L. icterohaemorrhagiae*)   23
Leptospira interrogans (*L. interrogans*)   23, 76, 302–3
Leptospira spp.   **74**, **302**
Leptotricia buccalis (*L. buccalis*)   23, 72
Listeria monocytogenes (*L. monocytogenes*)   11, **21**, 45, 234, **261**, 263, 264, **273**, **282**, 285, 286
Loa loa   144

Malassezia furfur (*M. furfur*)   **115**, 117, 210
Marburg virus   306
measles virus   98–9, **227**, **229**, 291–2
Melaninogenicus oralis   71
Micrococcus spp.   **21**
Microsporum audouini (*M. audouini*)   118
Microsporum canis (*M. canis*)   118, 119
Microsporum spp.   **115**
Moraxella catarrhalis (*M. catarrhalis*)   **49**, 50, 230, 231
morbillivirus (measles virus)   98–9, **227**, **229**, 291–2
Mucor spp.   114, 196, **229**
mumps virus   98, 99, 280, 291
Mycobacter pneumoniae   230
Mycobacteria spp.   330
Mycobacterium avium-intracellulare (*M. avium-intracellulare*)   78, 80, **81**, **300**, **309**
Mycobacterium bovis (*M. bovis*)   23, 78–9
Mycobacterium chelonei (*M. chelonei*)   23, **81**
Mycobacterium kansasii (*M. kansasii*)   23, 78, **81**
Mycobacterium leprae (*M. leprae*)   9, 23, 78, 80
Mycobacterium marinum (*M. marinum*)   78, **81**
Mycobacterium tuberculosis (*M. tuberculosis*)   23, 78–9, 155, **213**, 215, 232, 265, 280, **309**, 313, 315
Mycobacterium ulcerans (*M. ulcerans*)   **81**
Mycoplasma hominis (*M. hominis*)   86–7
Mycoplasma pneumoniae (*M. pneumoniae*)   2, 23, 86–7, 176, 226, 227, 229, **267**

Naegleria fowleri (*N. fowleri*)   122, **123**, 124, 266
Necator americanus   141–2
Neisseria gonorrhoeae (*N. gonorrhoeae*)   7, **22**, 48, **49**, 155, 168, 172, 215, 254, **254**, **273**, 287, 315
Neisseria lactamica (*N. lactamica*)   50
Neisseria meningitidis (*N. meningitidis*)   10, 20, **22**, 48, 49–50, 155, **173**, 220, 260–1, 262, **264**, **273**, 313, 326
Neisseria spp.   179, 180, 183, 187
Neisseria subflava (*N. subflava*)   50
Nocardia asteroides (*N. asteroides*)   **21**, 47
Nocardia braziliensis (*N. braziliensis*)   47
Nocardia spp.   46, 47
Norwalk virus   104

Onchocerca volvulus   144
Opisthorchis sinensis (*O. sinensis*)   **149**, 150

papillomavirus   199, **254**
Paracoccidioides brasiliensis (*P. brasiliensis*)   **116**, 120–1
Paragonimus westermani   **149**, 150–1
parainfluenza virus   98, 99, **173**, 221, 222, **227**, 230
paramyxovirus   98–9, 291
parvovirus   103–4, 215, **282**, 284
Pasteurella multocida (*P. multocida*)   **22**, **58**, 62, 207, **302**

Organism Index

Pasteurella spp. 184
Peptococcus spp. 21, 30, 37
Peptostreptococcus spp. 21, 30, 37
Plasmodium falciparum (*P. falciparum*) 129, 130, 131, **267**
Plasmodium malariae (*P. malariae*) 128, **129, 130**, 131
Plasmodium ovale (*P. ovale*) 128, **130**, 131
Plasmodium spp. **123, 309**
Plasmodium vivax (*P. vivax*) 128, **130**, 131
Pneumocystis carinii **123**, 126–8, 170, 187, **229**, 230, **300, 309**
Pneumocystis spp. **173**
Pneumoniae spp. **173**
pneumovirus (respiratory syncytial virus) 98
poliovirus 95, **96**, 265, 267, **267**
poxvirus 306
Prevotella melaninogenicus 218
Prevotella spp. 71, **217**, 218, 258
Propionibacterium acnes (*P. acnes*) 21, 203
Proteus mirabilis (*P. mirabilis*) **22**, 56, 250, **251**
Proteus spp. 51, **52**, 244
Proteus vulgaris (*P. vulgaris*) 56
Pseudomonas aeruginosa (*P. aeruginosa*) 10, **22**, 67–8, 156, 179, 181, 185, 188, 207, 208, 211, 225, 229, 231, **251**, 286, 297, 298, 321
Pseudomonas cepacia (*P. cepacia*) 67, 68
Pseudomonas maltophilia see *Stenotrophomonas maltophilia*
Pseudomonas spp. 182, **213**, 276, 313

rabies virus 112–13, 305–6
respiratory syncytial virus 98, 201, 221, **227**, 230
rhabdoviridae 112
rhinovirus 96, 155, 221, 230
Rhizopus arrhizus **116**, 119
Rickettsia akari (*R. akari*) 86
Rickettsia prowazeki (*R. prowazeki*) 86
Rickettsia rickettsii (*R. rickettsii*) 86
Rickettsia spp. 85–6, 305
Rickettsia tsutsugamushi (*R. tsutsugamushi*) 86
Rickettsia typhi (*R. typhi*) 86
rotavirus 104–5, **173**, 241–2
rubella virus 99–100, 282–3
rububavirus 98

Salmonella enteritidis 53
Salmonella paratyphi **52**, 53, 54, 55, 236
Salmonella spp. 13, 51, 171, 172, **213**, 271, **302**, 330
Salmonella typhi **52**, 53, 54, 55, 236, **309**, 324, 326
Salmonella typhimurium (*S. typhimurium*) **22**, 53
Schistosoma haematobium (*S. haematobium*) 151, **251**
Schistosoma japonicum (*S. japonicum*) 151
Schistosoma mansoni (*S. mansoni*) 151
Schistosoma spp. **149**
Serratia marcescens (*S. marcescens*) **22**
Serratia spp. 51, **52**, 53, 229, 250
Shigella boydii (*S. boydii*) 55, 235
Shigella dysenteriae (*S. dysenteriae*) 55, 56, 235
Shigella flexneri (*S. flexneri*) 55, 235
Shigella sonnei (*S. sonnei*) **22**, 54, 55, 234, 235
Shigella spp. 51, **52**, 54, 55, 315
small round virus 104, 242
Spirillum minor 306
Staphylococcus albus see *Staphylococcus epidermidis*
Staphylococcus aureus (*S. aureus*) 2, 20, **21**, 25–8, 93, 155, 157, 179, 180, 181, 184, 202, 203, 205, 207, 208, 209, 210, **213**, 214, 218, 220, 225, 228, **234**, 240, 245, 268, **272, 278**, 280, 288, 297, 298, **300**, 308, 313, 314–15, 319, 321
Staphylococcus epidermidis (*S. epidermidis*) 21, 25, **26**, 28,
154, 184, 214, **251**, 264–5, **272, 278, 279**, 280, 287, 297, **298**, 308, 321
Staphylococcus saprophyticus (*S. saprophyticus*) 21, 25, **26**, 29, 250, **251**
Stenotrophomonas maltophilia 68
Streptobacillus moniliformis (*S. moniliformis*) 70, **302**, 306
Streptococcus bovis (*S. bovis*) 35
Streptococcus epidermidis 6
Streptococcus faecalis (*S. faecalis*) **21**, 31
Streptococcus milleri (*S. milleri*) 36, 202, 245, 268, **272**, 308
Streptococcus mitior **278**
Streptococcus mitis (*S. mitis*) 35
Streptococcus mutans (*S. mutans*) 35, 216
Streptococcus pneumoniae (*S. pneumoniae*) 2, 5, 16, 20, **21**, 30, 34–5, 155, 156, **213**, 215, 220, 225, 227, 229, 231, **261**, 262, **264**, **272**, 280, 287, 288, 308, 326
Streptococcus pyogenes (*S. pyogenes*) **21**, 31–3, 180, **213**, 222–3, 290, 297, **298**, 308
Streptococcus salivarius (*S. salivarius*) 35, 217
Streptococcus sanguis (*S. sanguis*) 35, 216, **217**, **278**
Streptococcus spp. **173**, 187
Strongyloides stercoralis 142, **300**

Taenia saginata (*T. saginata*) **145**, 146–7
Taenia solium (*T. solium*) 145–6
togavirus **267**, 294
Toxocara canis (*T. canis*) 137, **138**, 140, 211, **302**
Toxocara cati (*T. cati*) 137, **138**, 140, **302**
Toxoplasma gondii **123**, 126, *127*, 211, **267, 282**, 283–4, **299, 302**, 307
Treponema carateum (*T. carateum*) 73, **74**, 75
Treponema pallidum (*T. pallidum*) 9, 23, 73–5, 179, **254**, 256–7, **267, 282**, 285
Treponema pertenue (*T. pertenue*) 23, 73, **74**, 75
Treponema vincenti 218
Trichinella spiralis **138**, 142–3
Trichomonas vaginalis **123**, 131, 132, 188, **251, 254**, 255, 256
Trichophyton mentagrophytes (*T. mentagrophytes*) 118
Trichophyton rubrum (*T. rubrum*) 118
Trichophyton spp. 114, **115**
Trichuris trichiura **138**, 141
Trypanosoma brucei gambiense (*T. brucei gambiense*) 132–3
Trypanosoma brucei rhodesiense (*T. brucei rhodesiense*) 133
Trypanosoma cruzi (*T. cruzi*) 133–4
Trypanosoma spp. **123**, 131, 132–4

Ureaplasma urealyticum (*U. urealyticum*) 23, 86–7, 255

varicella-zoster virus 101–2, 198, 199, 210, **227, 229, 267, 282**, 284, 292–3, **300**
Veillonella spp. **22, 217**
Vibrio cholerae (*V. cholerae*) 11, **22**, 65–6, 157, 237–8, 324
Vibrio parahaemolyticus (*V. parahaemolyticus*) **22**, 66, **234**, 240

Xenopsylla cheopsis 303

Yersinia enterocolitica (*Y. enterocolitica*) **22**, 56, 240
Yersinia pestis (*Y. pestis*) 56, 303–4, 324
Yersinia pseudotuberculosis (*Y. pseudotuberculosis*) 56
Yersinia spp. **52**, 215

Zygomycetes spp. 114